John Eric Erichsen

On Hospitalism and the Causes of Death After Operations

John Eric Erichsen

On Hospitalism and the Causes of Death After Operations

ISBN/EAN: 9783337162061

Printed in Europe, USA, Canada, Australia, Japan

Cover: Foto ©berggeist007 / pixelio.de

More available books at **www.hansebooks.com**

HOSPITALISM

LONDON: PRINTED BY
SPOTTISWOODE AND CO., NEW-STREET SQUARE
AND PARLIAMENT STREET

ON HOSPITALISM

AND THE

CAUSES OF DEATH AFTER OPERATIONS

BY

JOHN ERIC ERICHSEN

FELLOW OF THE ROYAL COLLEGE OF SURGEONS; SENIOR SURGEON TO UNIVERSITY
COLLEGE HOSPITAL; HOLME PROFESSOR OF CLINICAL SURGERY
IN UNIVERSITY COLLEGE; ETC.

LONDON
LONGMANS, GREEN, AND CO.
1874

NOTICE.

THESE LECTURES were delivered to the Surgical Clinical Class, at University College Hospital, during the past Winter. They were published at the time in some of the medical journals.

They have now been re-arranged, considerably extended, and in some respects altered. But the colloquial style of the Lecture has been preserved, as being better suited for the subject under discussion than the more didactic one of the Essay.

They are published in a separate form on the recommendation of friends whose judgment is as much respected as their approbation is valued by the Author.

JOHN ERIC ERICHSEN.

LONDON. *April* 1874.

CONTENTS.

LECTURE I.
ON THE RATE OF MORTALITY AFTER AMPUTATIONS.

General Propositions—Success in Treatment not kept pace with Progress in Art of Surgery—No Diminutions in Rate of Hospital Mortality—Amputations in University College Hospital—Table of—Simpson's Statistics—Parisian, American, and German Statistics—Billroth, Phillips, Lawrie, &c.—So-called 'runs of luck'—Causes of, threefold—Death after Operation not to be Confounded with Death from Operation—Table of Amputations in four London Hospitals that give Causes of Death—Hospital Mortality still very high—No Improvement during last Forty Years . . PAGE 1

LECTURE II.
ON THE CAUSES OF DEATH AFTER OPERATIONS.

Conclusions from last Lecture—Causes of Death fourfold—Shock and Septic Diseases differently felt in different Operations—Septic Diseases more frequent in some than in others—Influence of Shock—Influence of Septic Diseases—Hospitalism—Meaning of Term—Septic Disease in Hospital and Private Practice—Author's Experience—Influence on Mortality: 1. Of Maternity Charities—Lefort's Statistics; 2. Of Ovariotomy—Spencer Wells' Views; 3. Of Amputations in Active War—Franco-German War—Difference between Rate of Mortality in Huts and in Hospitals—Simpson's Statistics—Three Conditions may influence result independently of Hospitalism, viz. 1. Condition of Patient before Operation; 2. Skill of Operator; 3. After-treatment—Antiseptics 26

LECTURE III.

ON THE MODE OF PRODUCTION OF HOSPITALISM.

Evils of Overcrowding—Nature, not mere Number of Cases Important—Experience in University College Hospital—Contamination of Hospital Air—Tyndall on Dust and Disease—Organic Dust in Air—Impurities in Air we Breathe—Mode of Infection of Wounds by Organic Dust—Parkes, Rattray, Cunningham, Billroth, and Farr—Four Forms of Septic Disease—Hospital Gangrene—Erysipelas—Septicæmia, Pyæmia: Definition of these—Mode of Development of Pyæmia and of Erysipelas—Dissecting Students spread Infection—Semelweiss at Vienna Maternity . . . . . PAGE 59

LECTURE IV.

ON THE PREVENTION OF HOSPITALISM.

Overcrowding—how it Produces Septic Disease, by Local Infection and by Blood-poisoning—Three Conditions that chiefly influence Rate of Mortality in Hospitals: 1. Size; 2. Work done in it; 3. Mode of Construction—Farr's and Simpson's Views as to Effect of Size—Amount of Work done out of Proportion to Size and Arrangement of Hospitals—Varies greatly in different Hospitals—Table of Amputations in Proportion to Number of Beds in different Hospitals—Table of Operations during Three Years in University College Hospital—Construction of Hospitals—Evils of Present Plan—Lincoln Hospital—Opinions of Surgeons—Norfolk and Norwich Hospital—Views of Cadge—Impurity of Hospital Air arises from three Main Causes: 1. Domestic Offices; 2. The Out-patient Department; 3. Post-mortem Rooms and Erysipelas Wards—Some Points connected with Hospital Hygiene—Conclusion 84

ON HOSPITALISM

AND THE

CAUSES OF DEATH AFTER OPERATIONS.

LECTURE I.

ON THE RATE OF MORTALITY AFTER AMPUTATIONS.

THERE are few subjects in surgery of greater interest, and certainly there is none of greater importance, than that which relates to the cause of death after an operation, more especially when that operation is one that does not necessarily affect a vital part; for on its consideration necessarily depends the success, in its ultimate issue, of any operative procedure that we may undertake. On reviewing the recent progress and present state of surgery in reference to this grave question, I think that we are justified in coming to the following conclusions:—

1. That surgery, in its mechanical and manipulative processes, in its Art, in fact, is approaching, if it has not already attained to, something like finality of perfection.

2. That the Science of surgery has not advanced proportionately with the Art.

3. That the results of all operations, but especially

of those by which life is imperilled, are by no means so satisfactory as they should be; that the skill in the performance has far outstripped the success in the result; that the mortality is excessive after some operations, especially amputations; that it has not diminished of late years; that this mortality is particularly great in hospital practice, and is probably dependent on removable causes, and hence might be materially reduced, if these could be determined and their occurrence prevented.

The first two propositions I illustrated at some length in the Introductory Address [1] that I delivered at University College, at the opening of Session 1873-74. To the third, my time would not allow me to do more than advert; nor, indeed, had time allowed, would the delivery of an introductory address have been a fitting occasion to discuss so wide and important a question as this. The subject, however, is so interesting and so important, that I gladly take an early opportunity to direct your attention to it, to bring forward some facts in support of the statements I have made, and to give you the impressions that I have formed from a study of the whole question.

And on this subject I feel that I may speak with some degree of experience at least, for from the earliest period of my professional studies my attention has been directed to it. More than thirty years ago papers of mine on 'Congestive Pneumonia, consequent on Surgical Operations, Injuries, and Diseases,' were published in the 'Medical Gazette,'[2] and in the 'Medico-Chirurgical

[1] 'Modern Surgery: its Progress and Tendencies.' Lewis, 1873.
[2] 'Medical Gazette,' vol. xxvii. p. 794.

Transactions;'[1] and in the subsequent year another paper, on the 'Pathology and Cause of Death after Burns,' appeared in the 'Medical Gazette.'[2] And since that time the causes of death after operations have never ceased to occupy my anxious attention and much of my thoughts as a Hospital Surgeon.

In that past pathological era, and indeed long subsequently to it, surgeons, though recognising various visceral lesions in the form of congestions, inflammations, and abscesses in connexion with and consequent on surgical operations and injuries, failed to connect the two conditions together in the relation of effect and cause through the intermediate link of blood-poisoning; and in fact 'septic' diseases were but little known or regarded in their relation to surgical practice, and in their influence on its results, at the period to which I am now referring.

There is probably no collateral branch of knowledge that has a closer and more direct bearing upon the improvement of surgical practice, so far as the lessening of mortality after operations is concerned, than *hygiene*; and, if I do not greatly err, it is in this direction that we ought to look for some of the greatest improvements in modern surgery. Hygiene has indeed a double relation to surgery: it may be considered in its application to the prevention of diseases and deformities that render surgical interference necessary; and in its influence on the results of such interference or operation.

The first of these questions does not concern us here.

---

[1] 'Transactions of the Royal Medical and Chirurgical Society,' vol. xxvi. 1843.

[2] 'Medical Gazette,' Nos. 789, 790. 1843.

It is solely with the second that we shall have to deal; and indeed it is by its influence on the results of operations, rather than to their prevention, that the applicacation of hygiene to surgical science and practice has been, and will be, attended by the most important consequences.

The success of an operation has to be considered from two points of view—in its mere performance, and in the ultimate result it is destined to accomplish, whether that be the relief of the patient from some mere local malady or the preservation of his life from an otherwise incurable injury or disease. It may succeed brilliantly in one of these respects—in its mere perforanmce; it may fail miserably in the accomplishment of the desired end. 'L'opération, comme opération a réussie mais—le malade est mort,' has been observed with as much acuteness as severity by a French surgical critic. It is not with the successful performances of the operation itself that we have to do here; it is with the causes of its failure so far as the preservation of life is concerned. It is the study of those circumstances which, independently of the mere manipulative skill of the operator, influence the results of his operations, and often counterbalance all the good that the most advanced art, wielded by the most consummate skill, can effect, that is so important to the practitioner.

And here I do not speak of the mere local results. So far as they are concerned, there is but little to be desired. The results of most plastic, conservative, and ophthalmic operations are as satisfactory as the most sanguine could hope or the most critical expect. So also with respect to that multitude of minor operations that are practised for the relief of various distressing maladies, and which are followed by the happiest consequences. But when we come to consider the issues of those greater and graver operations by which the life of a patient is directly imperilled, we are constrained to admit that success in the result has lagged far behind and borne no relation to perfection in the execution of the operation, and that in this respect the highly polished Art of modern surgery far outshines its Science. But success in the result is, after all, the thing to aim at, and no amount of manual dexterity can compensate for its want. Dexterity, diligently as it should be cultivated, and highly as it must ever be valued, is only one element of success; and however important it is to be dexterous operators, it is better still to be successful ones.

We have, as I have elsewhere shown, carried the art of surgery to the highest degree of perfection of which, as an art, it is susceptible. But although we have undoubtedly immensely improved on the rapidity, the precision, and the simplicity of our operations, we are constrained to admit that we have not succeeded in rendering them proportionately less fatal. Here the surgeon has a wide field open before him in the future; and I can truly say there is no direction in which it can be cultivated that promises a more fruitful harvest than in endeavouring to make the success of the result balance the skill in the performance of an operation.

For it is useless, worse than useless, to ignore, and it would be reprehensible to deny, the fact that the mortality resulting from or consequent upon the greater operations has not only not diminished of late years, but has, there is reason to believe, in some cases actually increased. The present death-rate after lithotomy—even when making allowance for the application of lithotrity to the more favourable cases—is quite as great as it was in the days of Cheselden or of the great Norwich surgeons. Herniotomy is at least as fatal as it was in the hands of Hey and of Cooper. And the mortality consequent on amputations—the operations on which we possess the most extended statistics—has certainly not decreased, but if anything rather been on the increase, since Phillips and Lawrie published their tables. The significance of this fact is very great, for as the general mortality of the metropolis and of many other large towns has, during the last third of a century, been very decidedly on the decrease, whilst that of the hospitals situated in those towns has been stationary, possibly even increasing, it follows that the sanitary improvement in hospitals has not kept pace with that of the towns in which they are situated, and that thus there has actually been a decided increase in the ratio of their death-rate, when we compare it with the lessened mortality of the surrounding population.

Since July 1, 1870, up to December 1, 1873, there have been in University College Hospital eighty major amputations of the limbs, without including many minor and partial ones of the hand and foot. These cases have been taken as they came; they have not been in any way selected. None have been rejected in which it has been supposed that, by operating, the patient's prospects of recovery would, in any way, be improved. Every case has been most faithfully recorded by the surgical registrar, Mr. Beck, and they have either been published, or are in the course of publication, in the statistical records of the hospital

TABLE A.—*Table of 80 consecutive Cases of Amputation (excluding all partial amputations of hand and foot) performed in University College Hospital, from July 1, 1870, to Dec. 1, 1873.*

Amputation	Total	Cured	Died	Cause of death
PRIMARY.				
Hip-joint	1	0	1	Shock.
Thigh (above middle)	2	0	2	Shock in both.
Thigh (below middle)	3	2	1	Pyæmia.
Knee	1	0	1	Erysipelas.
Leg (upper half)	2	1	1	Exhaustion third day.
Leg (lower half)	5	4	1	Pyæmia.
Foot	2	2	0	
Arm	3	2	1	Died of internal injuries in a few hours.
Forearm	3	3	0	
Multiple amputations	3	2	1	Exhaustion.
Total primary	25	16	9	36 per cent.
SECONDARY.				
Thigh (below middle)	4	3	1	Exhaustion fifth day.
Knee	1	1	0	
Leg	2	2	0	
Shoulder-joint	1	0	1	Pyæmia.
Forearm	3	1	2	Pyæmia; tetanus.
Total secondary	11	7	4	36·3 per cent.
FOR DISEASE.				
Hip-joint	2	2	0	
Thigh	14	12	2	1 died of Bright's disease. 1 pyæmia & amyloid liver.
Knee and condyles	6	4	2	Pyæmia in both.
Leg	7	6	1	Pyæmia.
Foot and ankle	4	4	0	
Shoulder-joint	1	1	0	
Arm	5	4	1	Pyæmia.
Forearm	5	3	2	Exhaustion in both cases; old people, with destruction of wrist and cellulitis.
Total for disease	44	36	8	18·1 per cent.
Total of all cases	80	59	21	26·2 per cent.[1]

[1] Or, if we omit the case in which death resulted from internal injuries, and not from the operation, 25·2 per cent.

Of these eighty amputations, twenty-one have died, being in the proportion of about twenty-six per cent. of the whole. Although eighty cases of amputation is far too small a number upon which to found any general deductions, I may take these as a fair specimen of the general results of amputations that have occurred in this hospital since it was opened thirty-eight years ago.

I have been in the habit of publishing, in the successive editions of the 'Science and Art of Surgery,' the results of all the amputations that have been performed in my wards. They amount to a grand total of 307 cases; of these, 79 have died, being a mortality on the whole of as nearly as possible 25 per cent. The rate of mortality in the amputations in my wards in University College Hospital has been very uniform for the last twenty years. Thus, up to 1857 (2nd edition), it was 23·5 per cent.; in 1864 (4th edition) it had risen slightly, being 24·3 per cent.; in 1869 (5th edition) it had fallen to 24 per cent. The present rate again shows a slight increase. If we add the 80 cases in Table A to the 307 just referred to, we obtain a total of 387 major amputations at University College Hospital, with exactly 100 deaths, or a mortality of 25·8 per cent. These are continuous unselected cases. They include all that have been done in my wards and in those of Mr. Liston, from the foundation of the hospital to the present time (a period of thirty-eight years), and all that have been done in the hospital since July 1870; but they do not include the amputations done by the other surgeons between 1848 and 1870, of which I have no record. As they stand, I believe I am correct in saying that no other series of amputations of equal number, extending over an equal length of time, has ever been published in this country with an equally low rate of mortality. These amputations are primary and secondary—for injury and for disease; but include no partial ones of the hand or foot. As the mortality of the eighty cases that I shall take as my text corresponds, as nearly as possible, with the general mortality of the whole preceding series of amputations, I think they may be considered to be in all respects a fair specimen of these operations; and I prefer using them for our present purpose, on account of the care with which the causes that have led to the death of twenty-one of the patients have been worked out by the surgical registrar. For, although the general numbers that I have above given are as nearly as possible, if not absolutely correct, I have not sufficiently accurate data of the causes that led to the death of the 79 out of the 307 cases for statistical purposes. Now, a general mortality for many years of from 24 to 26 per cent. in all major amputations of the limbs for all cases, may be considered as a very satisfactory result, although there can be no question that it is one that admits, and that ought to be susceptible, of very great improvement. If we compare it with the results that have been obtained elsewhere, it is one of which we need not be ashamed.

Sir James Simpson, as is well known, has collected an enormous mass of statistics in connection with the subject of amputation-mortality.

These statistics are divisible into two distinct and separate sets of figures: one referring to amputations in town hospitals; the other to the results of amputations in country and private practice. The accuracy of these latter has been seriously impugned. With that question I have nothing to do at present. The accuracy of those relating to hospital practice has been admitted by all—even by his most determined opponents—for they have been derived from statistical returns furnished to him by the surgeons and registrars of the various hospitals to which they relate. The mortality attendant upon these is 1 in 2·4, or 41·6 per cent. The lowest mortality in any of the metropolitan hospitals referred to by Sir James Simpson is 34·4 per cent.; the highest, 47·3 per cent. These statistics are carried down to the year 1868; they may consequently be looked upon as modern. They are defective in one important respect, that the causes of death are not recorded. I have collected from various sources others which will show that there has been, with one exception, no material improvement since Simpson's were published, and that the same, or indeed a much higher, rate of mortality prevails in some other countries than in this. Thus, for instance, if we take the published records given in the reports of four of the largest London hospitals, containing together nearly 2,000 beds (Table B), carried down as late as last year, we find that out of a gross number of 621 amputations, 239 have died, or 36·7 per cent. (*vide* Table B); at the Edinburgh Infirmary, 43·3 per cent.; and at the Glasgow Infirmary, 39·1 (Simpson). The American returns are very good: thus, I find that at the Pennsylvania Hospital, the reports of which have been most accurately kept for a long series of years—from 1831 to

1860—the average mortality is very low, being only 24·3 per cent. (Norris); whilst at the Massachusetts General Hospital, Boston, out of 692 amputations, there were only 180 deaths, or a mortality of 26 per cent. (Chadwick). These figures correspond, as nearly as possible, with those of University College Hospital.

The mortality in the Parisian hospitals, on the other hand, is far greater than that in London. As given by Malgaigne and Husson, Holmes and Bristowe, it amounts to about 60 per cent. Lefort states that in Paris, from 1836 to 1863, out of 682 amputation cases, 397, or 58·8 per cent., died.[1] Billroth, when at Zurich, performed 163 amputations between the years 1860 and 1867. Of these 75 died, or 46 per cent. At Vienna, in 1868 his amputation mortality amounted to 43·4 per cent., but in 1869-70 it had fallen to 26·08. The number of the cases was, however, so small—only 23 in each series—that much importance cannot be attached to this.[2] Thus it will be seen that a mortality of from 25 to 30 per cent. is considerably below the average of most metropolitan hospitals in this country, and far below that on the Continent. In military practice the recent experience derived from the results of operations on the wounded in the great and destructive wars of modern times, on both sides of the Atlantic, is very unfavourable. But to these I need do little more than allude, as the disturbing influences at work during the progress of active war are so great and often so special, that they remove these cases into a category of their own,

---

[1] 'Malgaigne: Médecine Opératoire,' by Lefort. Paris, 1874. P. 499.
[2] Billroth, 'Chirurgische Klinik.' Zurich, 1866-67; Do. Wien, 1868; Do. Wien, 1869-70 (Berlin).

entirely apart from those of civil life. I may, however, mention that amputations of the thigh for gunshot injury were fatal, in the British army in the Crimea, in the proportion of 64 per cent. In the American Civil War, in a like ratio. In the French army, in the Italian campaign of 1859, the deaths amounted to 76, and in the Crimea to 90 per cent. (Lefort.)

In the consideration of the various questions connected with hospital mortality, as in all other matters that are alone determinable by vital statistics, it is necessary to attend to certain general conditions. In the first place, it is necessary that the numbers from which deductions are made should be sufficiently large not to be influenced by accidental circumstances, and next that they should extend over a sufficiently long space of time. The fact is well known to all surgeons, that surgical practice is peculiarly liable to what is often as incorrectly as irreverently termed 'runs of luck,' good or bad. Thus it is well known that in lithotomy, or in herniotomy, there may be long runs of successful or of unsuccessful cases in the practice of the same surgeon or in the same institution. If a surgeon has had a long run of success, he may think that he has at last mastered the secret of operating with certainty in any given disease. If, however, he go on long enough, the 'Nemesis of numbers' will certainly punish him for his presumption, and the wave of success on which he has been triumphantly carried will at last break and land him on the barren shore of the usual average. The man who boasts that he never loses a case after this or the other operation works at last for averages. He thinks more of his own credit than of his patient's good. He rejects, as unfit, those cases that are not likely to be successful, and reminds one of that sportsman who always hit his bird by taking good care never to shoot at one that there was a possibility of missing. These so-called 'runs of luck' have been well illustrated by the history of amputations. Lawrie[1] makes the remarkable statement, that at the Glasgow Infirmary, between the years 1794 and 1810, there were 30 amputations for disease of which 29 recovered; but on taking the central numbers from the register, he found that they showed a proportion of 11 deaths to 19 cures; and of the last 30, 22 recovered and 8 died. The general percentage of deaths after amputation in this institution was 39·1 per cent. In the Edinburgh Infirmary, of the first 99 cases, according to Simpson, only 8 died. At the Hospital D. (Table B), there was in 1870-71 a run of 18 major amputations without a single death; whilst in 1867, '68, and '69 there had been a total of 25, with a mortality of 12. Simpson gives the details of two remarkable runs of success at St. Bartholomew's Hospital, during which two periods there were in all 49 amputations, with only two deaths; yet, on continuing the record up to 1868, the average mortality amounted to 1 in 2·9. In the Pennsylvania Hospital, in the years 1838-9, of 24 amputations, out of which 11 were primary, only 1 died. At the Worcester Hospital there had occurred at one time 30 primary amputations in succession without a single death. Carden, operating with the long flap, had 17 consecutive cases of amputation of the thigh without a single death; but of the next 14, he lost 5. These 'runs of luck,' then, have

[1] 'Medical Gazette,' 1840.

not been uncommon in amputations. The reverse will also sometimes happen. I see by the reports of one Hospital D., that in one year all the amputations except two died in that institution, whilst fourteen out of fifteen cases of strangulated hernia recovered after operation. It would be as unfair to deduce any inference with regard to the average mortality of amputations in that hospital from such an instance, as of the average result of operations for strangulated hernia from the run of success that I have just mentioned.

The causes of these runs of success appear to me not to have been investigated with the care they deserve. The run of success is usually attributed by the operator himself to superior skill or care—by his colleagues and friends to 'chance.' Is it 'skill,' is it 'chance,' is it a 'special providence'? I cannot say which of the two latter it may be. It cannot be the former; for these varying runs of success will occur in the practice of every surgeon who practises long enough in public for others besides himself to form an estimate of his success or failure. When I use the word 'chance' in this matter, I employ it in the same way that we say that a man is killed by a 'chance shot' in his first action, whilst his fellow, 'more lucky than he,' goes through twenty unscathed. And doubtless in surgical operations, just as in games of chance, we may have a long series of uniform events developing themselves as a consequence of conditions that are in themselves most uncertain. It is, no doubt, quite *possible* that eleven operations out of a series of twelve may succeed or may fail, on the same principle that, if a shilling be tossed a dozen times, it may come down heads or tails for eleven times in succession. But, admitting the possibility of 'chance' in determining the result of operations—and I dislike much to use that word in so momentous a question as this, in which the life or death of a human being is concerned—I believe that, in order to explain these runs of success, we must, in the majority of cases, go somewhat farther and deeper into the consideration of the causes that determine events.

One great cause in many cases is, I believe, the prevalence or absence of epidemic septic influences at any given period. There are in all hospitals healthy and unhealthy seasons and periods—times in which every operative case, of whatever kind, does badly, from the influence of certain injurious epidemic conditions; others in which epidemics that are adverse to the success of operations are absent. To this I shall advert more fully in a subsequent lecture; I merely state the fact now.

But, besides this, there is another condition that influences materially these runs of success: I mean the personal care and supervision exercised by the surgeon. This is especially the case when new methods of operation or of treatment are introduced, in the success of which the originator takes a special personal interest. Thus, we commonly find that new methods of amputation—as by long or rectangular flaps, for instance—have been attended by a run of success in the early cases, which have probably been carefully selected and personally attended to by the surgeons who, having introduced the methods, were peculiarly anxious about their success. We find the same in the early cases of any new treatment, to which great attention is paid by an enthusiastic originator. The cases succeed, not because the method of operation or of treatment is superior to any that has preceded it, but because the surgeon who is personally interested in its success devotes more time and more attention to his cases than those do who take less interest in the method. And it is easy to understand how a possible combination of these three causes—chance, absence of epidemics, and personal care—may lead to a long run of that success which is most wrongly ascribed to 'luck.' Now, in order not to be misled by these runs of success, and with the view of neutralising the erroneous influences that they might exercise on statistical returns, it is necessary to employ large numbers of cases from which to draw our conclusions. But, in consequence of this very necessity of employing large numbers, it becomes imperative, in an enquiry like this, to go beyond the possible experience of any one man, and to draw one's deductions from reliable results that may be obtained from the published records of large hospitals. These will be found in Table B.

Next in importance to large numbers comes, undoubtedly, a similarity of the condition either in the patient operated upon or in the operation that is practised. Thus, it is not fair to compare the results of the same operation at different periods of life. This is well illustrated in the history of lithotomy, in which, as is familiarly known to every surgeon, the mortality increases almost proportionately to the age of the patient. So, also, it is impossible, with any degree of accuracy or of utility, to compare the results of dissimilar operations with one another; as, for instance, of amputations with lithotomy, herniotomy, or ovariotomy. So, also, we must not compare the results of operations on individuals in a very dissimilar condition of life, or in different hospitals the mortality of which is habitually very dissimilar, or even in the same hospital in different years, or under the influence of different epidemic seasons.

In all probability, Ovariotomy presents more uniformity in respect to the conditions under which it is performed than any other operation in surgery. Hence the statistics that we can obtain from it are of greater value in reference to the amount of mortality under the varying conditions in which the patient is placed, or in reference to the causes that lead to death, than can be furnished by an examination of statistical results of any other operation; for in ovariotomy there is more than a similarity, there is absolute identity, not only in the seat of operation, but in the nature of the disease and in the sex of the patient.

In amputations, on the contrary, there is in all probability more variety than in any other operation in surgery. There are not only those varieties that are common to most, dependent on age, sex, constitution, and condition of the patient; but there are the varying conditions that are inseparable from dissimilarity of cause for which the operation is practised. Thus, it may be done for injury, and that injury may be of any extent, from the perforation of a joint by a pistol-ball or its puncture by a needle, to the complete smash of a limb by a railway-buffer or cannon-ball. It may be done at any period after the infliction of the injury, within an hour or two, or not for months. Then, again, when amputations are practised for disease, that disease may either be simple or malignant, may be scrofulous or cancerous, may affect the bones, or the joints, or the soft parts.

In determining the mortality after operations there is a very important distinction to be made. It is one that has not been sufficiently attended to, and indeed it is one that, in the present state of surgical statistics, we are scarcely able to determine; I mean, to distinguish those cases of death that occur after any operation from causes and conditions in no way dependent upon the operation itself, from those that are the direct and immediate result of the operative procedure; in fact, to distinguish the accidental sequence—the *post hoc*—from the direct effect—the *propter hoc*. Until this is done surgical statistics will not be serviceable as guides in practice. When massed together, as is now the case, they give us the averages which are not applicable to individual cases. When disentangled, in the way I mention, they would give us data from which we might draw our inferences as to probable results in particular cases.[1]

Let me give you some examples of my meaning. A man is admitted into the hospital with a railway smash of one leg; amputation is performed; he dies collapsed, of 'shock,' if you will, in six hours; after death a laceration of the liver is found, the hemorrhage from which had led to his death. Now, here the patient had died *after* the amputation, but not *from* it. The operation was at least unnecessary; he would have died

---

[1] See Verneuil: 'De quelques Réformes à introduire dans la Statistique Chirurgicale, &c.: Archives Générales de Medicine.' 1873.

equally had it not been performed. A man is cut for stone; he dies in a few days; the kidneys are found in a state of disease that, though incompatible with life, had been undetected by the surgeon. Here the patient dies of fatal kidney disease, his death being at most only hastened by the operation.

Now, the cases that stand in the second category, that are directly due to the operation, and are not mere accidental sequences, but direct effects, are such as arise from tetanus, secondary hemorrhage, and septic diseases. A patient has a small tumour on the scalp, in no way detrimental to health; it is removed, and he dies in ten days from erysipelas. Another has a chronically diseased knee, in no way incompatible with life; it is expedient to remove the useless limb; pyæmia supervenes and death occurs. In fact, deaths after operations of all and every kind from septic diseases, and more especially from erysipelas and pyæmia, may be considered as being directly and immediately due to and the result of the operation itself—occasioned by causes that would have been inoperative if no operation had been performed. Hence, although 'shock,' as a cause of death after primary amputations, may require qualification and detailed explanation, a return of 'pyæmia' after an amputation, whether traumatic or pathological, requires none, and may be looked upon as the direct consequence of the operation—a disease that would not have occurred had the limb not been removed.

TABLE B.—*Summary of Results of Amputations in four Metropolitan Hospitals, in which Deaths from Pyæmia and Shock are recorded.*

Hospital and Years	Total	Died	Pyæmia	Shock	Percentage of Deaths
PRIMARY AMPUTATIONS.					
A. 1866–70	21	13	4	6	62
B. 1861–72	140	67	24	8	46·4
C. 1869–70	18	8	2	5	44·3
D. 1867–71	8	2	0	2	37·5
Total primary	187	90	30	21	48·6
SECONDARY AMPUTATIONS.					
A. 1866–70	20	16	9	2	80
B. 1861–72	53	30	10	1	56·5
C. 1869–70	5	1	1	0	20
D. 1867–71	6	3	2	0	50
Total secondaries	84	50	22	3	59·5
FOR DISEASE.					
A. 1866–70	89	31	15	0	34·9
B. 1861–72	215	56	16	0	26
C. 1869–70	17	5	0	0	29·5
D. 1867–71	29	6	3	0	20·6
Total for disease	350	98	34	0	27·4
Total of all amputations	631	239	86	24	37·8

As it is very far from my wish or intention to institute anything like a comparison between the results of operations in different hospitals, I have omitted the names of the institutions mentioned in Table B, and have designated them by letters. I ought, however, to observe that these four hospitals contain amongst them above 1,800 beds; and that, taken with University College Hospital, they together constitute about one-

AMPUTATION—THE TEST-OPERATION. 21

half of the accommodation in the eleven general London hospitals. They are the only hospitals in the reports of which the *causes* of death after operations have to my knowledge been published, except in the case of St. Bartholomew's Hospital, for the recent statistics of which I would refer to the Reports of that hospital. It is but right to observe, that in this great institution the death-rate after amputation has, since the year 1869, been lower than that of any other hospital with the statistics of which I am acquainted, and that the amount of hospital disease appears to be extremely small; but from the way in which the returns have been published, I have not been able to make out with accuracy the exact amount of this and other causes of death.

Although amputations have, on account of the frequency of their performance, the uniformity of skill with which they are practised, and the readiness with which their results are determinable, usually been taken as the test-operation, from the result of which the surgical salubrity of a hospital is determined and statistical deductions can be most easily made, yet they are, in point of fact, those operations from which it is most difficult to arrive at accurate conclusions with absolute certainty, unless the numbers employed are so large as to equalise the varying and disturbing conditions under which the operation is performed. In determining, therefore, the results of amputations, and indeed of all other operations, it becomes necessary, above all, to employ large numbers, and the experience or practice of any one man is scarcely adequate to furnish these.

As, however, we have by far a greater mass of statistics in connection with amputations than with any other operation in surgery, it is, notwithstanding the objections just stated, more easy to discuss some of the principal causes of death after operation in connection with amputations than with any other surgical procedure; and this is the more easy, inasmuch as the operation of amputation does not, in itself, interest or implicate any necessarily vital part of the body beyond the arteries which are severed, and which are now always safely secured.

It is a little curious and not uninstructive to trace these accumulations of amputation—statistics which have, during the last thirty years, grown to a gross total of very many thousands; and in doing so we are enabled to form a comparative estimate of the rate of mortality after these operations at the present and in former days.

Statistics of amputations were in this country first furnished to the profession by the army surgeons, who published the results of the experience obtained in the long French war, at the end of the last and early part of this century; and Guthrie's statistics of the result of amputations after gunshot injuries, in the Peninsular war, continue to be amongst the most valuable and interesting that we possess. At a later period, Sir Rutherford Alcock, the late Minister in Japan, published the results which he obtained when Surgeon-in-Chief to General Evans's army in the Carlist war of 1835 and following years. The first statistics in civil surgery that were published in this country were collected by Benjamin Phillips, who, in 1837, read a paper on the subject before the Royal Medical and Chirurgical Society.

Amongst other conclusions at which Phillips arrived was this, that the mortality after amputation of the thigh, deduced from a body of statistics collected in England, France, Germany, and America, amounted to about two in five cases; and that the average mortality after all amputations, taken collectively, was 23·7 per cent. These and similar conclusions, which subsequent experience has proved to be decidedly under the mark, were considered by the Council of the Society to be so untrustworthy that they refused the publication of the paper in the 'Transactions,' on the ground that some source of error must have crept into the statistics which showed so unfavourable a result. Phillips subsequently published more extended statistical enquiries, the result of which was that the average mortality after amputations amounted to 35·4 per cent., or 5·6 per cent. less than Simpson's estimate in 1868—viz. 41 per cent. as the mortality in hospital practice. His example was followed by Lawrie of Glasgow, who, in the 'Medical Gazette' for 1840, published very extended statistics of the amputations in the Glasgow Infirmary, showing a mortality of 36 per cent. The surgeons of the Royal Infirmary at Liverpool, and of the Northern Hospital, in the same town, followed his example, which was then taken up and very generally adopted throughout the profession. The largest collection of statistics that have been made on this subject will be found in the first volume of Cooper's 'Surgical Dictionary,' published in 1861. They were collected by J. Lane; and in Simpson's works are carried down to 1868.[1] The accuracy of these last statistics is admitted by all; the numbers being derived from official sources in all cases. The result from them is that, taking all amputations of all the four limbs in eleven of the largest general hospitals of this country, there is an average mortality, as has already been stated, of 41·6 per cent. In the report of Simpson's statistics those of University College Hospital are omitted. The way in which this omission occurred is as follows, and ought to be mentioned, as it is a proof of the care with which these statistics were compiled. When Simpson was constructing his tables he wrote to me requesting to be furnished with a return of amputations from our hospital. I replied that all the amputations that had been performed in my wards had been published in successive editions of my work on 'The Science and Art of Surgery,' and that I was unable to furnish him with any statistics from the practice of my colleagues. He very properly declined to publish a partial return from one institution, whilst he was giving full returns from all the others; and consequently omitted all mention of University College Hospital.

From the preceding observations, then, I think we are fairly justified in concluding: 1. That the rate of mortality after amputations in hospital practice still continues to be very high; 2. That it has not decreased since the publication of the first amputation statistics in civil practice about thirty-five years ago.

This fact, then, is certain, and it is as melancholy as it is true and incontestable, that, taking the average

[1] 'Simpson's Works,' vol. ii. pp. 280–400; Art. Hospitalism.

mortality after all amputations of the four limbs in the largest hospitals, in the hands of men of the most consummate skill in the great centres of civilisation, we come to this result, that the mortality calculated on large numbers varies from 35 to 50 per cent., but is steady and unvarying between these figures. This is a result that is but little creditable to surgery; and in some amputations, as of the thigh and at the hip-joint for injury, the mortality rises to the frightful and astounding height of from 60 to 90 per cent. In fact, so constantly do these numbers come out in hospital and army returns, that surgeons have almost come to regard them as representing the necessary, or (so to speak) the normal rate of mortality after amputations.

But is this really so? Must hospital surgeons ever remain content with losing from one-third to one-half of *all* their amputation cases, and nine-tenths of some? Is this frightful death-rate the necessary result of the operation, and thus beyond the control of our science and the skill with which our art is exercised; or is it dependent on causes that are preventable, and which may be counteracted or removed? Surely here is ample scope for science to aid the operations of our art, and to supplement it where it ceases to be any longer efficient.

LECTURE II

ON THE CAUSES OF DEATH AFTER OPERATIONS.

In the last Lecture I pointed out the percentage of mortality that still prevails after the major amputations. The figures that I laid before you admit of no doubt or cavil, as they have been drawn from the statistical reports which are now annually published by the officers of many of our hospitals; their accuracy cannot, therefore, be questioned. They are sufficiently large, and extend over a sufficiently long space of years, to equalise the varying conditions mentioned in the last lecture as tending to invalidate surgical statistics. They prove incontestably that the average mortality after amputations in general hospitals in this country, taken as a whole, is from 35 to 40 per cent., whilst the available Continental returns show a much higher rate. I also pointed out the important fact, that there has been no diminution in this rate of mortality during the last thirty-five years; that it is, in fact, higher than that furnished by statistics then published. If the figures are correct—and, for the reason given, I have no doubt of their accuracy—the deductions I have made from them are legitimate.

We will next proceed to enquire into the most important question of all, viz. the causes that have led to the production, and that are still leading to the perpetuation, of so unsatisfactory a result.

These causes may, by a reference to any published statistical table of Hospital Reports, be found to arrange themselves under four distinct heads. First, there are certain conditions inherent in the operation itself which dispose to, or directly determine, a fatal result; as, for instance, the exposure of the membranes of the brain in trephining, the opening of the peritoneum in operations for hernia, or the deep cellular planes of the pelvis in lithotomy. To these conditions it suffices to advert, and no description of them is rendered necessary, as they explain themselves. But it is well to bear in mind that the influence of such direct conditions as these is much increased by the complication of septic agencies.

Secondly, we have a series of causes which exercise a very minute influence upon the general rate of mortality, although they are individually serious and important; such, for instance, as tetanus, secondary hemorrhage, &c.

But undoubtedly shock and septic diseases are the two principal causes that determine death after the greater operations, such as amputations, resections, ovariotomy, herniotomy, lithotomy, &c.

The influence of shock and of septic disease is very differently felt in different operations. The greater the portion of body or of diseased structure that is removed the more severely is shock felt. In these cases, also, the influence of septic agencies becomes more marked. This is owing to two causes: 1. That the depression of the nervous system consequent on the shock, and on the loss of blood that is the frequent accompaniment of a great operation, tends to lower the resisting power of the system to all noxious influences, and thus predisposes to septic absorption, constitutional or local; and, 2. By the large surface of wound exposed, rendering local contamination more likely to occur. Hence operations which, though very important in themselves, do not entail the infliction of extensive wounds, such as those for the ligature of arteries, &c., are not likely to be attended by evil after-consequences either from shock or septic disease. The prolongation of the time expended in the performance of an operation also exercises an injurious effect by the proportional exhaustion induced, and consequent vital depression.

The influence of shock and septic disease in amputations is well-marked. Out of the 80 cases occurring at University College Hospital, which form the basis of these observations, I find that there were 3 deaths from shock (all primary), and 10 from pyæmia and erysipelas; leaving only 8 deaths to be accounted for by exhaustion and the other minor and more varied causes that I have mentioned.

On referring to Table B we shall find that of a total of 631 amputations, 110 died from shock and pyæmia together, or 17·5 per cent. of the whole operated on; whilst of the 239 deaths, 48 per cent. were from the combined influence of these two causes—and this is irrespective of those that are reported as dying of 'exhaustion,' which is closely allied to shock, or from 'erysipelas,' 'low cellulitis,' and forms of septic disease other than pyæmia. This terrible disease proved fatal in as nearly as possible 36 per cent. of all the deaths, and shock in about 10 per cent. of the deaths, or in 3·8 per cent. of all amputations.

But the respective influences of these two great causes of death after amputations will be found not only to vary greatly, according as the operation is primary, secondary, or for disease, but also to exercise very different degrees of influence in different hospitals, as may be seen by Table B.

Shock was most felt in primary amputations, in the proportion of 25 per cent. of the deaths; was but little fatal in secondary amputations, 6 per cent.; and was entirely absent as a cause of death in pathological amputations.

Pyæmia was fatal in about one-third, or 33 per cent., of the primary amputations; in 44·4 per cent. of the secondary; and in those for disease it again acquired nearly the level of the primary—viz. 34·6 per cent.

I find that pyæmia is proportionately more fatal after amputations of the upper than of the lower extremity, occasioning about 40 per cent. of the deaths in the former, against 34 per cent. of the latter, after amputations for all causes. In primary amputations the disparity is more marked, being about 50 per cent. of all deaths in the upper, against about 32 in the lower limb; shock, on the other hand, being more fatal by far after primary amputations of the lower than of the upper extremity, owing doubtless to the larger mass removed.

We shall proceed to consider these two conditions more in detail; and, first, with regard to shock.

The influence of shock is necessarily most felt in

primary amputations. Indeed, its fatal results are almost entirely confined to amputations performed within twenty-four hours of the infliction of the injury. I have never known a case of intermediate or secondary amputation, or amputation for disease, in which the patient died from this cause. Fatal shock, in fact, is the result of the combined depressing influence of the injury and of the operation. It occurs in the exact proportion of the severity of the injury, the amount of loss of blood, and the age of the patient. It is often rather referable to the injury than to the operation; and it becomes a question whether, in many cases of serious and almost hopeless smash of a limb, it might not be better to let the patient expire in peace, than subject him to the repetition of a shock which his nervous system will be utterly unable to endure. This is more especially the case in extensive crush and disorganisation of the lower extremity up to or above the middle of the thigh, such as are not unfrequent at the present day from railway accidents, in which the mangling of the limb rather resembles that produced by cannon-shot than by an ordinary injury of civil life. In these cases amputation through the upper third of the thigh, or at the hip-joint, is the only available operation. It is usually done in such cases. But is it ever successful in the full-grown adult? That is a question which deserves the serious consideration of hospital surgeons. I am not acquainted with a single case in which such an operation has succeeded in general hospital practice, in men who have arrived at full maturity. In children and young adults it has proved successful. The three cases in which it was done, out of the eighty University College cases, all died of shock. The same catastrophe has happened in every other case on record with which I am acquainted. It is an operation that has been abandoned by military surgeons in cases of compound comminuted fracture of the femur from bullet-wound in this situation; ought it not to be equally discontinued by civil surgeons in those more hopeless cases of utter smash of the limb that occur in their practice? For my own part, I shall never again, except in children and young people, amputate in that situation for such injuries—hopeless alike, whether left or subjected to the knife; but surely better for the patient to be left to die in peace than to be again tortured by amputation, which all experience has shown to be useless.

It is of importance to observe, in reference to these cases of death from shock after primary amputations, that the fatal result happens a few hours, usually within twenty-four, of the performance of the operation. Hence, although it may be disposed to by the previous condition of the patient, and the influence exercised upon his powers of endurance by the severity of the injury, the loss of blood, his age, &c.—for death from shock necessarily occurs more frequently under similar conditions of injury at advanced than at early periods of life—or even by season of year, yet it cannot in any way be affected by the conditions to which the patient is exposed subsequently to the performance of the operation, so far at least as hospital or other external influences are concerned. We must, therefore, look upon death from shock as a part of the general accident to which the patient has been exposed and of the injury that he has sustained, aggravated, doubtless, by the further depressing influence exercised by so serious an operation as an amputation possibly high up in one of the limbs. It is interesting to observe that season exercises an influence on the liability to death from shock after primary amputations. According to Hewson[1] of Philadelphia, it is most fatal in winter. The reason is obvious: the cold, to which the sufferer has been exposed at the time of the occurrence of the accident for which he has to undergo an amputation, is an additional cause of vital depression.

If, therefore, we want to improve our statistics of amputations—in other words, to lessen the mortality consequent on these operations—the first point to look to is not to amputate needlessly in hopeless cases of smash of the thigh high up, in order to give 'a last chance' to a patient whose vital powers have already been depressed to the lowest ebb by a fearful mutilation. Such amputations, which often consist in little more than the severance of a limb still attached to the trunk by shreds of muscle, ought scarcely to find their way into a statistical table professing to give the general results of operations the majority of which are more deliberately performed, and with a better prospect of success. They ought, in point of fact, to constitute a class of cases apart, in which the operation is subsidiary to the injury that has preceded and that leads to it, the more so as they are frequently complicated with internal injuries which are not detected until after the death of the patient.

Shock, as has already been shown, exercises its influence chiefly in primary amputations; far less in secondary ones; and disappears entirely, as a cause of death, in pathological amputations. There, however,

[1] 'Pennsylvania Hospital Reports,' 1869.

it is replaced by 'exhaustion.' This condition stands in the same relation to amputations for disease that shock does to those that are practised for injury. Death by 'exhaustion' or 'collapse' is, in fact, an indication that the impression produced by the operation on the nervous system has been greater than the already enfeebled powers of the patient were able to endure; and the frequent occurrence of these terms, as indicating the cause of death in any table of amputation statistics, may be taken as evidence of the operation having been practised too often, when the patient was already so enfeebled by long-continued disease, or so exhausted by discharges and suffering, as to be unable to support the additional depressing influence of a serious surgical operation. In the returns of Dr. Chadwick of the results of amputations in the Massachusetts General Hospital, Boston, 'exhaustion' is stated to be by far the most frequent cause of death. Out of the 180 fatal cases it is returned as the cause of death in 98 or 54·4 per cent.; whilst 'collapse' is stated to have been fatal in 21 or 11·8 per cent., and 'shock' in only 2 cases. I cannot but think that these terms are here used in a somewhat different sense to what we employ them, and that they rather represent what we should call 'shock.' However this may be, the fact is certain that in these Boston cases no less than 122 deaths out of 180, or 67 per cent., occurred from causes that were altogether independent of any septic hospital influence, a state of things that speaks highly for the sanitary condition of the hospital.

The next and by far the most important of all the causes of death after operations in hospitals is undoubtedly that exercised by the development of septic disease. The importance of this is obvious, from a consideration of the following statistics. Out of the 21 deaths that occurred in the 80 University College amputations (Table A), no less than 10, or nearly one-half, arose from this cause; and, out of a grand total of 239 deaths occurring in four metropolitan hospitals (Table B), I find that 86 died from pyæmia alone, without including other septic diseases and secondary septic visceral diseases; there being nearly four times as many deaths from pyæmia as from shock.

Dr. Chadwick, in the report of the Massachusetts General Hospital, returns 42 deaths from pyæmia out of the 180 fatal cases that occurred in 692 amputations of all kinds, being in the proportion of 23·3 per cent. on the deaths, and only of 5·7 per cent. on the whole number of amputations—a most remarkably low percentage of septic mortality, of which it may be taken as the whole and sole representative, there being no deaths returned from erysipelas or any other hospital disease.

The amount of septic disease varies very greatly in different hospitals, not only after amputations but after other operations and injuries. It is extremely difficult to form, not only a correct, but even an approximative estimate of the number of cases that occur, for many cases which are undoubtedly due to septic causes are returned under the heads of various visceral inflammations, as 'pneumonia,' 'pericarditis,' 'pleurisy with effusion,' &c. Besides this, there are certain forms of erysipelas or of diffuse septic inflammation which attack the peritoneum and cellular planes of the pelvis after herniotomy and lithotomy, and which are looked upon as cases of simple peritonitis or of infiltration of urine, when in reality they are diffuse erysipeloid inflammation.

In University College Hospital, during the three and a half years in which the 9 deaths from pyæmia after amputation occurred, we had 14 other cases of pyæmia, and 2 of septicæmia, all fatal; and 85 cases of erysipelas, of which 16 died. From this some estimate may be formed of the amount of septic disease that probably exists in those other hospitals in which the mortality from pyæmia alone after amputation far exceeds that which occurs in University College Hospital.

Now, it is the influences which lead to the generation of these various septic diseases, and not the diseases themselves, or any one of them, that have been termed 'Hospitalism,' a word originally introduced by Sir James Simpson, much objected to by some, possibly somewhat incorrect in its application, but yet explicit and convenient for our present purpose, and one to which, I think, no serious objection applies. We have long familiarly recognised the effect of hospital influences in speaking of 'hospital sore throat,' 'hospital gangrene,' and 'hospital plagues;' and I see no sufficient reason for abstaining from the use of a word that groups together all the deleterious influences met with in hospitals in one general term.

The word 'Hospitalism' has been objected to by some surgeons for whose opinions I entertain the highest respect; and if I venture to differ from these gentlemen it is because I still feel that the objections urged are not sufficiently valid to induce me to discard a term which is not only comprehensive and explicit, but which, in my view of the case, conveys a correct idea of the chief source of the influences its meaning includes.

But on this point I wish to be clear, and by 'Hospitalism' I mean, then, a septic influence capable of infecting a wound or of affecting the constitution injuriously. This septic influence—this miasma, or poison, if you will—commonly exists to a greater or less extent in all hospitals, or in any building, temporary or permanent, where large numbers of wounded and injured persons are congregated under one roof. It may be kept in check, even remain in abeyance, under ordinary circumstances, by close attention to hygienic measures; but will develop itself in proportion as these are relaxed; and it may at any time, at any season of the year, and under any circumstances, acquire extreme virulence, if the crowding together of the operated or injured having suppurating wounds be excessive —if, in point of fact, under such circumstances, one of two conditions be established: either that the cubic space of air for each patient be brought below a certain amount, or if ventilation be neglected, whatever that cubic space may be. But attention even to these precautions will not prevent the generation and development of septic influence, if the aggregation of the wounded be very large. And when once a hospital has become thoroughly impregnated by these influences for a length of time, no hygienic measures can restore it to purity and safety.

It is important to observe, that by hospitalism is not meant any one kind of disease that is peculiar to, that specially or only occurs in hospitals, and that never occurs, under any combination of circumstances, out of such institutions. By it is meant a general morbid condition of the building or of its atmosphere productive of disease, that, for the reasons just mentioned, is more rife in hospitals than elsewhere. Doubtless all the septic diseases that are met with in hospitals may be encountered in the practice of surgeons out of these institutions, but they are unquestionably infinitely more common in hospital than in private practice; and their causes are certainly different.

The question as to the relative frequency of septic diseases in hospital and in private practice is one that is extremely difficult of solution, and we possess as yet no definite data to guide us in the matter. The difficulty consists not only in instituting a comparison between patients so differently circumstanced before as well as after the operation, as is the case with those who are treated in and out of hospitals, but also leads in a very great measure to defining what is meant by ' pyæmia;' for it is on the occurrence or not of this disease in private practice that so much difference of opinion exists. About erysipelas there appears to be no contention, for although undoubtedly more common in hospitals than in private, and often conveyed from hospital to private patients, yet it is not uncommon out of hospitals; and it must not be forgotten that erysipelas, unlike acute pyæmia, is not necessarily a traumatic disease; it often arises from constitutional causes, from trivial local irritations, and largely from cold acting on individuals constitutionally disposed to it, and in these individuals it may readily follow on the infliction of a slight wound, whether accidental or surgical, in hospital or in private.

Erysipelas, in some form or another, then, is the septic disease that is most frequently met with in private practice; and undoubtedly very severe cases of this disease are occasionally seen under circumstances where they would scarcely be expected. I have frequently, in my practice, met with as serious cases of phlegmonous erysipelas, after often trivial injuries, as I have ever seen in hospital, although I have comparatively rarely met with serious forms of erysipelas after operations in private. But in these cases I have been unable to trace it to the influence of external causes. It has arisen in individuals of broken, gouty, or otherwise diseased constitutions, or of evil habits, rendering them liable to low inflammations. Persons, in fact, in whom erysipelas might occur spontaneously, or from cold, without the starting-point of a wound being needed.

Pyæmia in my experience is very rarely met with out of hospitals, and in hospitals—at least in that one where my opportunities of observation have occurred—we very rarely indeed meet with amongst, or admit a case from the out-patients; almost every instance having occurred in patients who have for some time been in the wards; and I can safely assert that during the last twenty-five years I have never seen a single case of acute pyæmia or septicæmia following an operation in private practice. I have met with cases of blood-poisoning of a sub-acute character, accompanied by abscess in various parts of the body—usually in the cellular tissue —as the result of self-infection in some cases of malignant pustule, dissection-wound, bed-sore, &c. The only instances in which I have seen metastatic pulmonary abscesses following pyæmia, independently of operation or external injury, have been in acute necrosis of bone, especially of the petrous portion of the temporal. There is undoubtedly a connection between acute inflammatory disease of bone, whether it accompany or end in necrosis, or assume the form of osteo-myelitis, that is especially liable to lead to pyæmia.

The cases of 'blood-poisoning,' with cellular or lymphatic abscess following the other forms of disease above mentioned, differ so completely in all their symptoms, in cause, and, above all, in fatality, from this acute traumatic pyæmia, that I cannot include them under the same category of disease; but must look upon them as distinct and dissimilar affections, though perhaps of the same type. In considering acute traumatic pyæmia as essentially a hospital disease, and in saying that I have never met with it out of an hospital after operations, I know that I am enunciating an opinion, and making a statement that differ widely from the views and experience of some distinguished surgeons. But if my opinions and experience differ from those of some, they certainly agree with those of others. I have lately been called to see patients in two large provincial towns widely removed from each other, each being furnished with a county hospital of old standing, containing from one hundred and forty to one hundred and fifty beds. The surgeons whom I met in consultation on these occasions were men of the highest experience and in very extensive operative practice. They both told me the same story—that they were much distressed by the large amount of pyæmia in their respective hospitals, which had become so serious, notwithstanding the careful and vigilant employment of sanitary measures, that it was a question with the governing bodies of each of these institutions whether the existing buildings should be pulled down and reconstructed at a very great cost. These gentlemen both told me that they had never seen a case of pyæmia out of the hospitals in the towns in which they were situated. One of them mentioned a single exception in a case of lithotomy. I think it is impossible, in the face of facts such as these, which might easily be greatly multiplied, to doubt that there is a special, if not specific, morbid influence, capable of being generated, and of continued persistence, in large hospitals.

But although I have been fortunate enough not to happen to have met with pyæmia after operations in private, I do not for a moment doubt the possibility of its occurrence. We have the evidence of many surgeons of great experience to the contrary, and indeed it is only reasonable to believe that such unfortunate cases may occur in several ways: by the inclusion within the system of septic or putrid clots and discharges, more especially in operations on the uterine or urinary organs; by some fault in the operation itself, as when a vein is transfixed and held open by a tied ligature—whether it be a femoral, hemorrhoidal, or spermatic vein; and, lastly, possibly if a combination of the same circumstances that may be met with in hospitals were to occur elsewhere, as in a close, overcrowded, badly-drained, or ill-ventilated house, the same condition of septic influence might be developed.

It necessarily follows that, if there be an increased rate of mortality occasioned by septic influences generated within hospitals, there must be a rate of mortality smaller in the direct ratio of that which is produced by these causes in those patients who are operated upon out of hospitals. In other words, the success attending operations in private ought to be greater than that which follows similar procedures in hospitals, in the direct proportion of the deaths that arise from so-called hospitalism, other conditions being equal. As an evidence of the proportionably small amount of septic disease that occurs out of, in comparison with that which is developed in, hospitals, it may be stated that Simpson found that, out of 160 deaths after amputation in private and country practice, only eight occurred from pyæmia and one from erysipelas. The majority of the deaths arose from shock. This part of the present enquiry is one the elucidation of which is involved in extreme difficulty, owing partly to the comparative absence of statistics of private operations, and in some degree to the different conditions of patients, so far as social status, habits of life, &c., are concerned, that are operated upon out of hospitals, in comparison to those that are the inmates of such institutions.

There are, however, certain classes of cases into which these disturbing influences do not intrude themselves, and between which comparisons may, therefore, more justly be made, than can be done in others. The cases to which I refer are met with—1, in maternity charities; 2, in the operations for diseases special to women; and, 3, in military practice.

Attention was first of all drawn to the high rate of mortality that occurs in hospitals in comparison to what is met with in private practice, by obstetric practitioners finding that the mortality attendant on parturition in maternity charities was infinitely greater than that which occurred amongst poor women delivered at their own homes. The statistics of Lefort[1] on the rate of mortality following the delivery of nearly two million women in different parts of Europe—one-half in their own homes and the others in lying-in hospitals—are so distinct and definite in their results as to leave no doubt whatever upon the infinite increase of risk that is attendant upon parturition in a lying-in hospital. Lefort, after discussing at length the rate of mortality amongst women confined in their own homes and in public institutions, sums up thus :—'With respect to hospitals and maternity charities it seems below the truth when I estimate the mortality only at one death for every thirty-two deliveries. . . . The figures present themselves thus to us :—

	Deliveries.	Deaths.	
At home	934,781	4,015, or	1 in 212
In hospital	888,312	30,594 „	32

'I can thus say that the mortality amongst women confined in clinical hospitals and maternities is out of all proportion to that which occurs in private.' Simpson, whose authority on such a subject as this cannot certainly be questioned, however much it may be impugned by some in more strictly surgical matters, cites these figures in order to show the greater danger to life which women undergo who are confined in lying-in hospitals over those who are delivered in their own homes.

It is well known that it is absolutely impossible

[1] 'Des Maternités,' par Dr. Leon Lefort. Paris, 1866, 4to.

to establish a maternity ward in a general hospital without exposing the women confined in it to the greatest possible peril of life; and in every instance, I believe, in which it has been attempted in London, the mortality has been so great that it has become necessary to close the ward. The fact is certain, that a woman has a better chance of recovery after delivery in the meanest, poorest hovel, than in the best-conducted general hospital, furnished with every appliance that can add to her comfort, and with the best skill that a metropolis can afford. The conditions of a parturient woman are not altogether dissimilar from those of one who has undergone a surgical operation; and a comparison between the two, although not absolutely accurate, is sufficiently so to entitle us to draw this conclusion, that there is a septic influence existing in hospitals which is highly destructive to the life of women who have recently been delivered.

But if we compare the results that have been obtained in and out of hospitals in that operation which may justly be looked upon as one of the greatest glories of modern British surgery—Ovariotomy—we shall find somewhat similar results; namely, that the mortality after this operation, when done in general hospitals, amounts to from 60 to 80 per cent., whilst in private, in the hands of Keith and Spencer Wells, it is only from 18 to 24 per cent., and, indeed is gradually on the decline.

Nothing is more interesting and instructive than the early history of Ovariotomy. It owes its origin and its establishment in practice entirely to the success that attended its performance in the hands of country and private practitioners. Almost all, if not all, the early successful cases were done on private and not on hospital patients. It was tried in the London hospitals; but so great was the mortality following the operation when there performed, that there was the greatest danger of its falling entirely into disrepute and neglect. The operation was denounced as unjustifiable, and the operators were stigmatised in opprobrious terms by two of the most eminent, and ranked amongst the boldest, of the operating surgeons of that day—Lawrence and Liston. It has never taken its place as an operation practicable like others in large metropolitan hospitals. It has been proved by a sad and disastrous experience that if ovariotomy be practised in a large hospital, and if the patient be placed in a general ward—or even if secluded in a private one, she be exposed to hospital influences—her chance of recovery is rendered so small that no prudent surgeon will now undertake the operation in such circumstances. In some general hospitals in which ovariotomy has been largely practised, a special building detached from the hospital has been erected for the reception of such cases, and every possible care has been taken to prevent the contamination of the patient by influences generated in, or emanating from, the parent institution. Now, the mortality after ovariotomy in general hospitals amounts to 76 per cent., whilst in private practice Spencer Wells at most has lost only 24 per cent., or less than one-third of the hospital rate of mortality. And this amount of loss is, with increasing aptitude and experience, actually still on the decline. Thus Keith of Edinburgh has achieved the marvellously successful return of only

27 deaths in 144 cases, or a mortality of only and about 18 per cent.

On such a question as this the statements made by so distinguished and successful an operator as Spencer Wells must be considered as authoritative. After pointing out the importance of isolation and of close attention to sanitary arrangements in these cases, he concludes with these words: 'And the question seriously presents itself, whether ovariotomy, or any other surgical operation attended with risk to life, should ever be performed in a large general hospital in a large town, except under such circumstances as would render removal to the country or to a suburban cottage hospital more dangerous.'[1] And he states further, that his conviction is, 'that the surgeon who hopes to obtain better results than have hitherto been obtained must place his patient, as nearly as possible, in the position of a person in a private house in a healthy situation.'[2]

That which holds good with ovariotomy must surely be equally applicable to other great operations, and would doubtless be found to be so, if their comparative statistics were worked out; and if the rate of mortality after Ovariotomy is more than three times as great in general hospitals as it is in small institutions and in private practice, a more or less correspondingly high rate of mortality may be supposed to attach itself to other of the great operations by which life is directly imperilled.

When we turn to the subject of amputations we find it somewhat difficult to institute so accurate a

---

[1] 'Diseases of Ovaries,' p. 326.
[2] 'Medico-Chirurgical Transactions,' vol. lvi.

comparison between cases requiring this operation in civil practice in which it is performed, in and out of hospitals.

But in military practice this difficulty does not exist. Amputations in active warfare are practically performed on individuals under very similar conditions, so far as age, sex, previous condition of patient, cause and nature of injury are concerned. The only differences to which the sufferers are exposed are the different conditions to which they may be subjected after the operation has been done. Now, on this point the recent experience of the Franco-German war confirms that which has been derived from the results of operations practised in all modern wars, namely, that those patients have the best chance of recovery who are treated in the open, under canvas, in temporary huts, with the slenderest possible shelter, perhaps; whilst the mortality from pyæmia, erysipelas, hospital gangrene, and other septic diseases is developed almost in the direct proportion of the accumulation of wounded in regularly constructed buildings, converted for the purpose into temporary hospitals. The employment of hut-hospitals in the field during the late war in France proved incontestably that septic hospital influences of a destructive character may be almost entirely prevented. These huts, which were single-roomed, built of rough weather-boarding, raised on stages about two feet from the ground, were isolated one from the other. The lightness and looseness of the construction adapted them admirably for ventilation, the wind blowing freely through on all sides. Mr. Berkeley Hill, who visited these hut-hospitals at Saarbrück, states that each hut contained about fifteen beds, which were almost all occupied; that there was a complete absence of anything like a sick-room smell in any of the hut-hospitals, whilst in buildings of every other kind, used for the reception of the wounded, it was easy to detect that peculiar odour so universal in Continental hospital-wards. He states:—' In all the hut hospitals I visited the condition of the patients was most satisfactory; not a single case of pyæmia had occurred. In the houses and other buildings converted into hospitals, in too many instances, hospital diseases had begun to appear.'

It is impossible to give a better illustration of the development of hospitalism under, apparently, the most favourable circumstances for its prevention, than by a reference to what occurred in the orange conservatories which were converted by the Duke of Hesse-Darmstadt into temporary hospitals for the reception of the wounded. These orange-houses, Mr. Hill states, were wide, lofty rooms, with windows on the southern side that reached from ceiling to floor, and opened freely. They were situated in a fine garden, and the patients treated in them had every appliance of comfort and cleanliness. Several cases of pyæmia had occurred in them, and one of extensive gangrene, which ceased as soon as the patient was removed from the orangery to a small tent; and the surgeons in charge became thoroughly impressed with the unfitness even of such a building as a conservatory for the accumulation of wounded. The same happened with the school-house at Saarbrück—a large building that had been occupied for a few weeks as a hospital—apparently well-adapted for this purpose, having lofty rooms, an airy staircase, large windows, and situated in the outskirts of the town; yet pyæmia and hospital gangrene had been so rife within its walls, that one ward had already been closed. These and similar instances, which I might multiply to a considerable extent, prove as clearly as it is possible to demonstrate any sanitary fact, that septic influence, or, in other words, hospitalism, becomes generated by the accumulation of wounded persons under one roof, and may be prevented by their isolation in scattered groups. It is only in this way—viz. by the absence of septic diseases amongst those operated on in the field, whilst they are fearfully destructive amongst the wounded who are lodged in hospital—that we can explain the remarkable statistics published by Guthrie, as to the results of primary and secondary amputations in the Peninsular war. Of 291 primary amputations done in the field, and necessarily treated in the open, only 24, or 1 in 12, died; whilst of 551 secondary amputations done in hospital, 265, or nearly one-half, died. This statement, which, I believe, has never been challenged, is of itself almost conclusive as to the comparative safety of the two methods of treating amputated soldiers. In fact, a military hospital in active warfare differs from a civil one in this, that every case in it is one of operation or severe injury; hence the liability to septic poisoning becomes very greatly increased from the great contamination of the air, and the emanations from wounds in all states of suppuration and disorganisation.

It is well known that Sir James Simpson has attempted to prove by a very large body of statistics, collected from hospitals on the one side, and from the records furnished by private practitioners in the country on the other, that amputations practised in hospitals are far more fatal than those performed on patients in their own homes. The results to which he arrived were as follows: that out of 2,089 cases of amputations in large hospitals in this country, 855, or 1 in 2·4, had died. The accuracy of this body of statistics is admitted by all the results having been furnished by officials connected with the various hospitals. Of 2,098 cases occurring in country and private practice, the deaths were returned as only being 226, or at the rate of 1 in 9·2. This body of statistics has not been accepted with the same implicit confidence that is attached to the former. It is doubtless possible that the figures may not be absolutely correct, and that certain as yet undiscovered sources of fallacy may have introduced themselves into his tables, but these fallacies have not as yet been pointed out; and although the figures have been objected to, they have certainly not been disproved.

It is clear that if these tables are incorrect the error in them must have proceeded from one of two sources: either from the medical men who furnished the data on which they were constructed, or from Simpson, who tabulated the figures furnished to him by fellow-practitioners. Now, it is surely impossible to believe that the different medical practitioners who furnished Simpson with these figures should have sent false returns; and yet, unless they have done so, or unless their returns were falsified by Simpson himself, a supposition that is as monstrous as it is incredible, these figures must be admitted to be accurate.

It is, however, of great importance to bear in mind

that whilst admitting the accuracy of Simpson's figures—and the more closely I have studied them the less I am disposed to doubt their accuracy—it is not equally necessary to admit the accuracy of the conclusions that he has drawn from them; for although he may be perfectly correct in stating that up to the year 1868 the mortality in general hospitals in this country, after the major amputations, was 1 in 2·4, whilst that of country and private practice was only 1 in 9·2; and whilst he may be perfectly correct in stating that out of 377 cases of amputation of the forearm, occurring in private and country practice, only two died, whilst of 244 occurring in hospitals, 40 died, he may not be equally accurate in the conclusions at which he arrives as to the causes of the difference in result between hospital and private practice, based upon these statistics, for undoubtedly the difference in the rate of mortality may have been dependent to a great extent upon the difference in the constitution and condition of the patients before they underwent the operation, rather than to the influences to which they were exposed after being subjected to it. In fact, we may use the figures, but discard the inferences drawn from them.

Setting aside, then, the conclusions to which Simpson arrived by an analysis of his own tables—which conclusions are undoubtedly open to the grave objections that have so forcibly been urged against them by Callander and Holmes—I will confine myself to the conclusions that may legitimately be drawn from the comparisons I have just instituted. From these we find that the influence of those conditions to which patients are subjected in hospitals exercises a most marked and a very decided effect upon the prospect of their recovery. We find that seven times as many women die after confinement in hospitals as out of them; that ovariotomy is from three to four times as fatal in general hospitals as in private practice or small institutions, in which special sanitary precautions can be taken; that in military practice the death-rate after amputations is out of all proportion greater when patients are placed in hospitals than when treated in the open or even in hut-hospitals.

With respect to the influence of hospital conditions on amputations in civil practice, this may with certainty be said: that in hospitals a high and generally a non-decreasing rate of mortality is maintained, and that a considerable percentage of the high death-rate is dependent on septic disease. To what extent this amputation-mortality in civil hospitals differs from or exceeds that which occurs in private practice, under conditions that are in all respects, except those of hospital influences, similar to those that occur in the hospitals, we have as yet no data to determine, for the conditions affecting patients who seek hospital aid are undoubtedly so different from those of patients treated in their own homes, in towns provided with hospitals, in many circumstances besides those that they encounter after admission into hospital, that we are not as yet in a position to come to any definite conclusion on this subject, and indeed may never be able to arrive at it.

In order to make anything like a just comparison between the results of hospital and of home mortality, it would be necessary to compare the results of operations as they affect the patients of hospitals and those of a class similar in all social respects who undergo the same kind of operations in their own homes. So far as town populations are concerned, this has never been done, and is scarcely possible, inasmuch as the poor in all large towns, suffering from injuries or affected by diseases that require operation, necessarily apply to hospitals for treatment and operation. Hence the only comparisons that are practicable are those between the poor inmates of general hospitals and the wealthier inhabitants of the towns in which they are situated, which would be clearly unfair, and must lead to false conclusions, as there are so many influences, in state of health, habits of life, &c., that would exercise a disturbing effect quite independently of the mere fact of residence in or out of hospital. So also the comparison between the results of operations in general hospitals and in country practice, as instituted by Simpson, is, for reasons that have already been adverted to, and that have been fully pointed out in the lengthy discussions that have taken place on his statistics, liable to error and to the deduction of incorrect inferences. Setting aside, however, all these considerations, there can be no doubt of this, that if the present high rate of mortality after amputations be not dependent on those influences to which the patient is subjected by his residence in hospital after the operation has been performed, and which are included under the general term Hospitalism, it must be natural and inherent in the operation, and cannot be altered, or it must be dependent on some condition unconnected alike with the operation or the hospital in which it is performed.

CAUSES OF HIGH MORTALITY. 53

That this rate of mortality is not natural to, or necessarily inherent in, the operation, is evident from the fact that it varies greatly in the same hospital, having, for instance, been reduced at St. Bartholomew's from 36·6 per cent., at which the amputation-mortality stood up to 1869, to about 18 per cent., to which low level it has now been brought—that it varies greatly in different hospitals under precisely similar circumstances, except so far as hospitalism is concerned, and that it differs widely in hospital and civil practice, and cannot thus be dependent merely on the operation itself.

If, then, this high and varying mortality be not necessarily dependent on, natural to, or inherent in the operation itself, it must be due either to circumstances resulting from the hospital influences to which the patient is exposed, or to one of the three following conditions to which he is subjected after its performance. Let us briefly examine these.

1. Can the condition of the patient's health, his social status, his occupation, exercise any influence in determining the present high rate of mortality? Upon this point I need say but little, as I have no intention to compare town with country, hospital with private patients, but to compare the results of one hospital with another in the same town. Now, I may take it for granted that the conditions of health are the same in the patients in all metropolitan hospitals; that there is no difference in health or in social status between the patients of a hospital in which the mortality amounts to 18 or 25 per cent. after all amputations and those of one in which it reaches nearly 50; and

yet these are the differences that are met with in different metropolitan hospitals. We find the same differences existing in the hospitals of Paris; and, on reference to Simpson's admitted statistics, it will be found that the hospitals of county towns differ also largely one from the other in the rate of operation-mortality. Without, therefore, comparing town and country patients, we find the utmost possible discrepancy in the results obtained from different institutions in the same town, in which the conditions of the patients must necessarily be as nearly as possible similar in all that respects habits of life, occupation, and resulting constitution.

2. As to the second condition that may be supposed to lead to this different rate of mortality in different hospitals—namely, any difference of skill that may exist—it is not to be entertained for a moment. There are, doubtless, many capital operations in surgery—such as lithotomy, the ligature of arteries, herniotomy, &c.—in which it may be supposed that different operators possess varying degrees of skill, which might influence the result; but in amputation, that simplest of all the great operations, there can be no difference in this respect; and, indeed, the results that are furnished by different hospitals bear no proportion whatever to the known skill and professional ability of the surgeons connected with them; and it is one of the greatest evils of the prevalence of septic disease in a hospital—call it hospitalism or not, as you please—that it neutralises the highest surgical skill, and renders abortive every care that can be bestowed upon the patient.

3. The third possible condition that may influence

results of operations, independently of hospitalism, is treatment. Its influence on the results may possibly be somewhat great, though on this important point we possess, as yet, no reliable evidence. It is a curious fact that the surgeons of the past generation seem to have been almost entirely ignorant of the existence of septic disease, and of its influence in modifying injuriously the results of surgical operations. There are, no doubt, scattered hints in the works of the more philosophical surgeons of the day bearing upon this point, but septic disease was but little regarded, and its more prominent features scarcely recognised. Surgeons thirty-five years ago were chiefly concerned with methods of operating and modes of treatment, and they referred the varying conditions of success or failure to the practice that was adopted in these respects. We find long discussions and controversies upon the comparative advantages of the flap and circular methods of amputation, of immediate and delayed union; but no reference to those more general causes that are now known to override all these comparatively minor considerations of fashioning of flaps or mode of procuring union of wound. At this period, indeed, the secondary effects of pyæmia and septicæmia in the visceral abscesses and congestions produced were recognised; but the intermediate link of the blood-poisoning was but little, if at all, suspected. But if we go from this past pathological era and come to more recent times, we shall still find that a great, and I cannot but think an undue, importance has been attached to peculiar methods of operation and to particular plans of treatment. Let me speak on this point solely from my own personal experience. The 307 amputation cases that have occurred in my wards have all, I believe, without a single exception, been done by the flap operation. The patients have been subjected to various methods of treatment. In the early periods, up to twenty-five years ago, Liston—and I, acting afterwards on his precepts—generally treated amputation wounds by leaving the flaps open, with a piece of wet lint interposed, but otherwise fully exposed to the air for from four to six hours, until all oozing had ceased and the cut surfaces had become glazed. The flaps were then brought together, a strip of water-dressing laid along the edge of the wound, and an attempt made to procure union by adhesion. I afterwards employed different methods of treatment, generally bringing the flaps together immediately after the operation was completed, and dressing the stump in the operating theatre, sometimes washing the surface with a solution of chloride of zinc, with alcoholised water, or carbolised solutions. But, whatever method of treatment was adopted, the mortality was, as nearly as possible, the same, ranging, as I have stated in the first lecture, from 23 to 25 per cent.; in fact, it is quite certain that no influence whatever has been exercised on the result in my practice by any method of local treatment that has been adopted.

Of the antiseptic treatment I can as yet say nothing positive; it has been tried in some cases in my wards, and with success, but not, as yet, in a sufficient number for me to come to any conclusion as to its utility in operation wounds. Of its great advantage in chronic abscesses I have seen enough to leave no doubt on my mind. Theoretically, 'the antiseptic method' is perfect. It fulfils all the requirements that can be desired in the management of a wound. It may be, and I believe it is, equally good in practice, but, as I have already said, this is a point yet to be determined. The essential points in the local treatment of any wound are, absolute rest, scrupulous attention to cleanliness, the absolute purity, so far as freedom from all decomposable organic matter is concerned, of everything that is brought into contact with it, be it air, or instruments, or dressings, or surgeon's fingers, and close personal supervision. In all these respects the 'antiseptic treatment' of Lister, and Callender's method of managing stumps, leave nothing to be desired; and, if I were to venture an opinion upon a subject which is still *sub judice*, I should say that it is in this that their great merit in practice consists; and indeed rest, cleanliness, isolation, and ventilation are the great points on which Callender lays, and justly, so much stress. But we have, as yet, to learn the real value of antiseptic methods of treatment; and this can only be done by the observation of a very extended series of cases in which these plans of treatment have been employed, and comparing the results thus obtained with an equally extensive set of cases treated by other methods under as nearly as possible the same conditions in the same hospital. It would clearly lead to an incorrect conclusion if we were to compare the mortality resulting from treatment by antiseptics and by other modes, at different seasons of the year, or in different years, when epidemic influences might be present or absent, or in two hospitals, the mortality of which for years past has been so dissimilar as is presented by those contained in Table B, as compared with University College, with a mortality of 25 per cent., or St. Bartholomew's, with one of 18 per cent. The only fair method of comparison is between the practice of one surgeon who uses the antiseptic method with that of another who does not, in the same institution and at the same time. This has never as yet been done; or, at least, if done, the results have never been made public; and, until we are in possession of such comparative returns, we can attach no importance to the publication of isolated cases, or even of short and partial runs of success which have, as stated in Lecture I., occurred after almost every new method, and to many surgeons.

But if the high rate of mortality at present existing after amputations and many other operations, dependent upon septic diseases, and which differs so widely in different hospitals in London, is not due either to the condition of the patient before the operation, to the skill with which it is practised, or to the treatment that is adopted, to what are we to refer it? Why, necessarily to the first of the four conditions; namely, to the hospital influences to which the patient is subjected after the operation. It is these influences that give rise to the septic diseases that are so fatal, pyæmia alone being the cause of death in more than one-third of the fatal cases; and, if this could be removed, we should be able to lessen our mortality proportionately. In other words, instead of having an average death-rate of about 37 per cent. after amputation, we should bring it down to 24 at most, if only the pyæmic cases could be saved; and considerably lower, if the mortality from other septic disease, such as erysipelas, could be done away with or materially lessened.

LECTURE III.

ON THE MODE OF PRODUCTION OF HOSPITALISM.

By 'overcrowding' a condition of atmosphere is produced which may, according to the condition of the individuals so congregated together, produce diseases of various types. The overcrowding of uninjured individuals will produce some variety of typhus: the 'gaol-fever' of a past generation was an instance of this kind.

The overcrowding of wounded people—whether the wound be accidental or surgical matters not—will develop septic disease in one of four forms, viz. hospital gangrene, septicæmia, pyæmia, or erysipelas.

It would lead me altogether away from the subject of these Lectures were I to enter into a discussion of the general questions of the primary origin of these diseases, for an attempt at its solution would lead us far astray into the wide field of speculation and of hypothesis; it would lead us into a discussion of those great and much-vexed questions of autogenesis and heterogenesis—of spontaneous or of germ generation —questions that have occupied the attention and the time, and which have exercised the ingenuity of scientific men of all countries and of all ages since the days of Aristotle; but questions which are as far from their solution now as when he wrote 'De Animalium Generatione.' For we have done little more than push the question back, by the aid of modern and improved means of investigation, from the mode of development of the mollusc, or maggot, or carrion-fly to that of the *Bateria*.

That the septic poison which, when once generated, impregnates a wound, and thus gives rise to hospital gangrene, pyæmia, or erysipelas, is capable of transmission through the medium of the atmosphere, is undoubted. That it may be generated by 'overcrowding' is equally certain. The results of the most extended and recent observations in the development of this scourge of hospitals in actual warfare, which have of late years been numerous, have left no doubt whatever on the minds of military surgeons that 'overcrowding' is the direct occasioning cause of the whole class of surgical septic diseases. But the word 'overcrowding' is liable to misconception, and requires explanation. It is very important to be clear on this point; for by it is not meant the mere heaping together of sick and wounded in one ward or building to an extent beyond what it is intended to hold, or only limited by its size; but by 'overcrowding,' in the sense in which it is here used as the cause of the generation of the septic poison, is meant the accumulation within one ward, or under one roof, of a greater number of patients than is compatible with such a degree of purity of air as to render the septic poison incapable of development; or, if generated, of propagation in it.

It would, in fact, appear that the air of a ward is capable of oxidising, destroying, or absorbing a certain amount of morbid emanations from the contained patients; but if these emanations be developed too rapidly or too abundantly the air becomes overcharged with septic matter, and then all the ill effects of 'overcrowding' at once develop themselves. The contamination of the air of a ward may thus take place even though the actual number of patients lying in it be below what it is constructed to hold, as readily as by the introduction of one single patient beyond the number that the ward is calculated to accommodate with safety. Thus it would be more correct to say that the special evil effects of overcrowding, and the special form they will assume, are rather dependent on the nature than on the actual number of such cases that are contained in any given ward or building at one time. There is no evidence that an accumulation of unwounded patients to any extent can develop hospital gangrene or pyæmia, whatever other diseases may thus be generated. Pyæmia could not have been generated in the 'Black Hole' at Calcutta; but a very trifling excess in the number of open or suppurating wounds, as must occur at times in every hospital, beyond what any given ward is capable of holding with safety, constitutes 'overcrowding' in a surgical sense, and will infallibly generate septic disease.

I can illustrate this observation from what has happened in this hospital. Ward 1, my male accident-ward, contains a cubic space of 21,924 feet. It is intended to hold fourteen patients, giving a cubic space of 1,566 feet to each, or, if we deduct for space occupied by furniture, &c., about 1,500 cubic feet. But not only is the cubic space sufficient, the floor-space is also ample. There is a space of about six feet between each bed, and ten feet between the opposite rows; the ward being sixty-three feet long by twenty-four feet wide. It can be well ventilated by a row of windows on each side, a fireplace at one end and on one side, and a large door, with an open air-hole above, communicating with the passages of the hospital at one end. In this ward we have had many outbreaks of septic disease, pyæmia, and erysipelas, invariably the result of the accidental accumulation within it, not always of too large a number of patients, but of too many severe cases of operation and of injury with wounds in a state of suppuration. The average number of open wounds in this ward is about seven, or half the number of patients it contains; and, if these wounds be not severe, such a number can be contained in it with moderate safety. But if they rise above this in number, or if the majority be severe, then septic disease will certainly break out. The outbreak in Ward 1, in November 1872, which will be fully described hereafter, was thus produced, the ward containing at the time a remarkably severe set of cases; in fact, the conclusion at which we have arrived is this, that if Ward 1 contain nine or more cases with open wounds, at least three of which are severe, as amputations or similar cases, an outbreak of septicism in some form is imminent. Thus, then, it is the nature of the cases, and not their number merely, that constitutes the danger of the development of surgical septic disease by so-called overcrowding; and doubtless epidemic influences favour the development of those septic outbreaks that are thus directly produced.

In what these epidemic influences consist I know not. So far as my experience and observation go, they are connected with very different meteorological conditions. In wet and dry weather, in heat and cold, during prevalence of easterly and westerly winds, they have equally manifested themselves. The existence of an epidemic condition, in addition to the influence of overcrowding, is evidenced by the admission from out of doors of erysipelas cases and allied diseases, or by their appearance amongst the out-patients in unusual numbers. But the influence of different meteorological conditions on the development of epidemic septic diseases has yet to be worked out. In reality we know little or nothing that is positive about it.

The influence of season of year is, however, somewhat better ascertained. Thus the observations of Dr. Hewson of Philadelphia, that have already been referred to, point to the winter months as those in which pyæmia is most rife—an observation which has long been in accordance with my own experience at University College Hospital, where we have always suffered much from septic disease in the surgical wards during the winter and early spring months. This often appears to be owing, however, less to the direct influence of season of year than to its indirect effect in preventing ventilation; it being difficult, in wet or cold weather, to get the nurses or patients to submit to having the windows opened so as to allow an influx of fresh air in a constant current, and to prevent the ward from being ventilated wholly by air drawn from the hospital corridors. The air-supply that is thus admitted is, therefore, as is usual in hospitals, already in a state of contamination, and liable to generate disease.

Now, what is this condition of contamination in

hospital air that causes the development of septic disease in a ward the cubic and floor space of which are sufficient, and which is kept scrupulously clean? That is the next point for enquiry. In the well-known, I may say celebrated, lecture of Professor Tyndall on 'Dust and Disease,'[1] that distinguished philosopher points out the important fact that the atmosphere under all circumstances contains floating particles of organic matter; that the motes which dance in the sunbeam are of this character. He states that the air of our London rooms is loaded with this organic dust, nor is the country air free from its pollution; and he goes on to observe that, however much it may disguise itself from ordinary observation, 'a powerful beam of electric light causes the air in which the dust is suspended to appear as a semi-solid rather than as a gas. Nobody could in the first instance without repugnance place the mouth at the illuminated focus of the electric beam and inhale the dirt revealed there. Nor is the disgust abolished by the reflection that, although we do not see the nastiness, we are churning it in our lungs every hour and minute of our lives. There is no respite to this contact with dirt; and the wonder is, not that we should from time to time suffer from its presence, but that so small a portion of it would appear to be deadly to man.'

Watts states that a minute quantity of organic matter—one grain in 200,000—is found in the purest mountain air. This increases in the country at a lower level, and in large towns reaches the condition stated by Tyndall in the above extract. Even in sea-

[1] 'Fraser's Magazine,' July 1870.

air, according to Dr. Rattray,[1] it is not absent. This organic matter is the result, in all cases, of vegetable and animal emanations, and it necessarily varies in its composition, according to the source from which it is derived. Rattray, in a most interesting paper in the volume referred to, gives an analysis of ship-air; and, after pointing out a variety of impurities of a gaseous character, states that the more solid impurities are of animal, vegetable, and mineral origin, from the skin, lungs, &c., of the crew; from the ship, as minute particles of wood, paint, cordage, whitewash, leather, &c.; from the bilge, containing occasionally microscopical animal and vegetable organisms; from the stores, particles of bread, cotton, wool, &c.; and others which I need not mention. He states that it is the volatile organic matter thrown off by the skin which gives ship-air its close and often nauseous smell. Parkes, in his admirable work on 'Practical Hygiene,' gives the following account of hospital air.[2] At page 88 he says: 'I have examined the air of various barracks and military hospitals, and have detected large quantities of epithelium from the skin, and perhaps the mouth.' And at page 99 he says: 'The most important class of disease produced by impurities in the atmosphere is certainly caused by the presence of organic matters floating in the air, since under this heading come all the specific diseases. The exact condition of the organic matter is unknown; whether it is in the form of impalpable particles, or moist or dried epithelium and pus-cells, is a point for future enquiry.'

[1] 'Medico-Chirurgical Transactions,' vol. lvi.
[2] 'Manual of Practical Hygiene,' 3rd ed. 1869.

And again, at page 106: 'The air of a sick ward, containing as it does an immense quantity of organic matter, is well known to be most injurious. . . . At a certain point of impurity erysipelas and hospital gangrene appear. The occurrence of either disease is, in fact, a condemnation of the sanitary condition of the ward.' And again, at page 100, he states that 'erysipelas and hospital gangrene in surgical wards are often carried by dirty sponges, dressings, &c. Another mode of transference is by the passage into the atmosphere of disintegrating pus-cells and putrefying organic particles; and hence the great effect of free ventilation in military ophthalmia, in erysipelas, and hospital gangrene.' And, lastly, Dr. Douglas Cunningham of Calcutta finds that the atmospheric dust largely consists of spores of fungi, and that the majority of these are living and capable of growth and development; and that bacterial matter exists also in dry dust. So that when this is added to putrescible fluids a rapid development of fungi and bacteria occurs. Billroth inclines to the same view, namely, as to the possibility of the diffusion of contagion by organic particles, for he says: 'I can entirely agree with the miasmatic origin of pyæmia, if by "miasma" is understood dust-like dried constituents of pus, and possibly also very small living organisms accompanying them, which, in badly ventilated sick-rooms, are suspended in the air, or adhere to the walls, bed-clothes, dressings, or carelessly cleaned instruments. These bodies are all pyrogenous when they enter the blood, and they will of course collect chiefly where there is the best opportunity for their development and attachment.' And further on he says: 'From my own experience I hold to the opinion that the infection of the whole body comes from the wound, whether the poison finds circumstances favourable to its development in the wound and surrounding parts, or whether it be introduced into the wound already developed.'[1]

Dr. Farr makes the following important remarks on the point: 'One great evil has often counterbalanced all the advantages (of hospitals). The collection of a number of persons exceeding those of an ordinary family under one roof has hitherto always had a tendency to increase the dangers of disease, for several diseases are, like fire and ferments, diffusible. The danger is increased when all the inmates are sick, for their breath and excretions spread through the wards. The dangers, too, are likely to increase in a faster ratio than the numbers.'[2]

Mr. Sympson makes the following remarks on this subject: 'It is not difficult to understand why disastrous outbreaks of pyæmia, erysipelas, and sloughing of wounds should occur under unfavourable hygienic conditions, if we do but realise the fact that man is continually poisoning the air around him by the worn-out materials of his body, which escape in the exhalations from his lungs and skin; that in disease these effete products are considerably increased in amount, and that in hospitals still further contamination of the atmosphere occurs by the addition of the effluvia from

---

[1] Billroth, 'General Surgical Pathology and Therapeutics,' translated by Dr. Hackley, p. 346.

[2] 'Twenty-fourth Annual Report of the Registrar-General.' 1863. P. 231.

sores and wounds. We all know how speedily a crowded, ill-ventilated room becomes close; and that the emanations from our bodies are easily putrescible our sense of smell assures us when we enter an unventilated apartment which has been occupied on the previous evening.'[1]

Here, then, we find, on the combined authority of the most able and recent observers, abundant evidence not only of the existence in the atmosphere of large quantities of suspended organic matter, but of animal *débris* and exfoliations, and of other organic particles, capable, under favourable circumstances, of generating septic diseases. We are, however, still in complete ignorance as to the precise nature of the septic poison that produces those various forms of disease which we recognise as being due to its influence. We know it, in fact, by its effects, but are ignorant of what it essentially consists. That there are different forms of septic poison appears more than probable. We find that the different varieties of septic disease are not interchangeable, but are as distinct in themselves—in the symptoms they present, in the course they run, in the pathological conditions that are found—as any of the zymotic diseases. In whatever way originating, and in whatever it may essentially consist, there can be little doubt that this septic virus is communicable from patient to patient through the medium of the organic particles of various kinds with which the atmosphere of any crowded building, be it hospital barrack or man-of-war, is invariably charged; and when we reflect on the exceedingly minute, infinitesimal—in fact, inappreciable—quantity of any animal virus, as that of

[1] 'Letters on the Lincoln County Hospital,' p. 7.

small-pox, cow-pock, or syphilis, that is needed to communicate its own disease when applied to a fitting soil, we can easily understand how the virus of the septic diseases that occur in surgical wards may be transmitted from wound to wound even on so slight a vehicle as organic atmospheric dust. Dr. Angus Smith truly says: 'If we measure size by percentage it will appear small; but still smaller will appear the strychnine that destroys, if we estimate the amount as a percentage of the weight of our bodies.'[1]

But the exposure of a wounded patient to an atmosphere rendered organically foul by overcrowding will dispose to the occurrence of pyæmia in another way, viz. by depressing his vital powers. The absorption of the impurities by which the air is polluted into his blood through the medium of lungs or skin, though probably not capable of producing pyæmia, will yet render him more susceptible to that local contamination of the wound in which it takes its true origin; for in the low state of health thus induced that plastic barrier which surrounds every granulating surface, and intervenes between it and the healthy tissues, and which Billroth has shown to serve as a barrier against the absorption of putrescent matters, becomes weakened and broken down. It is this plastic rampart that, stretching across the veins, constitutes the thrombosis, which so long as it exists unbroken bars the passage, and thus prevents the entry of septic matters into its interior; but when broken down, though but through a fissure, not only admits the transudation of putrescent fluids from the wound, but actually itself becomes con-

[1] 'Air and Rain,' by Dr. Angus Smith, p. 9.

verted into those flocculent and septic emboli which, carried into the circulation, give rise to stasis in distant organs, and form the centre of resulting multiple abscesses. These emboli, indeed, the result of disintegration and semi-puriform liquefaction of the preservative thrombus, act, when they gain entry into the capillaries, in exactly the same way in producing distant stasis as the suspended solid particles in ink did when injected into the veins of animals by Cruveilhier. Their septic condition then leads to a purulent centre in the midst of the capillary embolem, and thus the metastatic lung or other visceral abscess is produced. But it may reasonably be supposed that exposure to a polluted atmosphere, containing large quantities of organic matter in suspension or solution, may act injuriously in another way, viz. by preventing the elimination from the skin or lungs of those effete matters which would be thrown off from these surfaces in a purer atmosphere, but may be retained in the system under the influence of one already containing organic moisture and suspended particles in quantity sufficient to interfere with free transpiration.

There are four of these septic diseases thus produced universally recognised by surgeons—hospital gangrene, septicæmia, pyæmia, and erysipelas. Let us very briefly study the mode of development of these.

With regard to hospital gangrene I need say but little. It is recognised, by the concurrent testimony of all military surgeons, that this disease originates in the first instance as a direct consequence of the overcrowding of the wounded in hospitals that are insufficiently ventilated. The experience of the Franco-

GENERATION OF HOSPITAL GANGRENE. 71

German war added confirmation, if any were needed, to this view of the origin of this pestilence; which, however, when once generated, is capable of indefinite propagation by contact, through the medium of fingers, instruments, sponges, and surgical appliances. The occurrence of hospital gangrene in civil hospitals is now, fortunately, extremely rare; and its development in such institutions is of itself an evidence that the sanitary condition of the building is for the time at least in a bad state. The repeated recurrence of hospital gangrene in a civil hospital would undoubtedly be discreditable to those who had the management of its sanitary arrangements. In University College Hospital we have now had no outbreak of hospital gangrene for more than twenty years; and I trust never to see it here again, as having been developed within the walls of the hospital. It is altogether a preventable disease, and ought never to occur in an institution that is conducted on proper sanitary principles. The few cases that we have had of late years have been brought into the hospital from without; and, curiously enough, last summer two came to us—of a mild form, certainly —from a convalescent institution.

But the most important of the septic diseases are undoubtedly septicæmia, pyæmia, and erysipelas. These diseases are commonly looked upon as being more or less allied, and they may be so, in so far as their etiology is concerned,—in 'overcrowding,' and in epidemic influences that predispose to their occurrence, but in all other respects they differ widely, and are not interchangeable.

Acute pyæmia is essentially a hospital disease. As

has already been stated in the last Lecture, it seldom occurs except in hospital practice; and perhaps the best proof that we have of the rarity of the occurrence of true pyæmia out of hospitals exists in the rarity of the admission of a pyæmic patient into a hospital from without whilst labouring under the disease. Almost every case of pyæmia that I have seen in hospital practice has originated within the building itself. By a reference to Table B it will be seen that the amount of pyæmia varies very greatly in different London hospitals, which is an additional proof of its being dependent for its origin on conditions that are more rife in some hospitals than in others. So far as University College Hospital is concerned, there appears to be, at present, a tolerably uniform amount of it. In three years, from July 1870 to July 1873, we had twenty-three cases: nine in connection with amputations, fourteen with other operations and injuries. Of these twenty-three cases, four occurred in the latter half of 1870, eight in 1871, eight in 1872, and three in the first half of 1873. As to the mode of its development, there can be no doubt that it is the result of the exposure of wounds to an atmosphere overcharged with organic matter emanating from other patients with suppurating wounds. What I stated in the last lecture about the infection of the military hospitals with pyæmia during the Franco-German war, and the comparative freedom of the 'hut-hospitals' from this disease, points clearly to its cause.

It is very important to give some definition of these two diseases, septicæmia and pyæmia.

By 'septicæmia' I mean a blood disease, a form of

typhus or 'putrid fever,' directly occasioned by the absorption into the system of putrescent matters from fœtid ulcers, necrosing cancers, &c., which may thus become self-infecting. In it there are no rigors or sweats, but extreme depression of vital power, and usually rapid death, with typhoid symptoms. After death no metastatic abscesses are found. But the solid organs, more especially the spleen, the liver, and the lungs, are found darkly congested, loaded with blood, soft, and at times almost pulpy. It is a disease that may affect the uninjured as well as the wounded; and the reason why a person who has been the subject of a severe operation, or of a serious injury, is more liable to septicæmia than another, appears simply to be that his constitution has been weakened by the shock to the nervous system or by the loss of blood sustained, and that consequently he is rendered less resistant to the invasion of any disease of a miasmatic type.

The term 'pyæmia' is used in a very wide and elastic manner, and by many is made to include various forms of blood-poisoning. In these Lectures I only speak of one form of it, the true acute traumatic pyæmia, that form which Virchow has shown to be dependent upon venous thrombosis leading to embolism, and that embolism to metastatic abscesses. In order to produce true pyæmia the embolism must be of a putrid, puriform, or septic character. It is this form of pyæmia that is so common after all amputations or injuries implicating the bones, especially if followed by osteo-myrlitis. It is very rare, so far as my own personal experience goes it scarcely ever occurs, in operations or injuries merely affecting the soft parts, by which neither a large vein nor a cancellous bone is opened up. It would be altogether foreign to the object of these Lectures were I to enter into anything like a general description of pyæmia; but in order to avoid misunderstanding as to the meaning that I attach to the term, I think it well to say that I do not include under it those forms of blood-poisoning of a more ·chronic character which arise from self-infection, whether from abscess, urethral or uterine discharges, or other similar sources. Nor do I consider any case to be truly pyæmic where the deposit of pus is directly in the course of the absorbents leading from the part originally affected, and where no metastatic deposit has shown itself in distant parts or internal organs. We have yet to work out the real distinctions that exist between the true acute traumatic pyæmia and those chronic forms of diffused abscess in the areolar tissue which will develop themselves in certain conditions of the system after trivial injuries and operations, and which appear to be more allied to forms of furuncular than of true pyæmic disease, or those more remarkable forms of non-suppurating plastic deposits that occur after certain forms of blood-poisoning, and that are so often and so incorrectly assumed to be of a rheumatic nature. But the consideration of these questions I must leave to a future and different occasion.

Septicæmia is not of very common occurrence, yet there is a certain proportion of deaths arising from it. In this hospital we have about one death in the year from this cause. Its influence on the general mortality after operations is, therefore, but trifling, but its existence is an indication of a septic influence; and I believe that a certain number of cases of low or irritative fever, following operations and injuries, partake of the septicæmic type; in fact, the gradation from traumatic fever into septicæmia is easy, and, I believe, not very unfrequent. Surgeons themselves suffer at times from small whiffs of this poison. Who that has been long connected with a hospital has not at times, after the examination of a sloughing cancer or some other horribly putrescent case, felt feverish, depressed, prostrated for some hours, or a day or two, conscious of the absorption of a poisonous effluvium, which, after a period of febrile depression, eliminates itself from the system by an attack of offensive diarrhœa, the eruption of a pustule, or perhaps by some more distinctly localised inflammatory action, such as tonsillitis or a boil? The cause of septicæmia appears to me to be somewhat obscure. It does not appear to be distinctly connected with overcrowding, but rather with the development of putrescent discharges from unhealthy or malignant ulcers. The offensive discharge from ulcerated cancer uteri is supposed by some to tend very specially to its production, and it has been a cause of death in ovariotomy, when practised in the same building in which a woman suffering from this disease was lying.

I look upon pyæmia, when of traumatic origin, as being primarily the local septic impregnation of a wound by organic atmospheric matters, in a condition capable of developing change of such a character in the wound that its fluids decompose, its surface becomes foul or sloughy, and the veins leading from it become plugged with soft clots, putrid or easily decomposable. The constitutional symptoms of this dread disease—the prostrating rigors, the profuse transudations from skin and lung, saturating the bedclothes and contaminating the air around with a faint sickly odour, the high temperature, the extreme mental depression—are all consequent on the entry into the circulation of the septic virus deposited from air on wounds, absorbed into the veins, and thence transmitted through the system.

Pyæmia stands next to hospital gangrene amongst the septic diseases of local origin. It is less marked in its local phenomena; it is far more developed in its constitutional symptoms. But, though less marked locally, it is most distinctly characterised. Mr. Beck, who has made the post-mortem examinations of the twenty-one cases of pyæmia that have occurred at University College Hospital during the past three years, with a degree of minuteness and care that leaves nothing unnoted, states that in every case except one —a case of necrosis of the tibia, in which there was no open wound until a few hours before death—distinct local evidences were found. In fourteen cases the local origin of the embolism was evidenced by broken down and putrid clots in the veins leading from the part; and in six there were foul sloughy wounds.

It is important to observe that in pyæmia the venous thrombosis whence the fatal emboli proceed does not always exist in the veins leading directly from the primary seat of operation or disease; but some secondary or accidentally developed condition may lead to it. Thus, in a recent case of amputation of the thigh that died of pyæmia in University College Hospital, the veins of the stump were quite free and healthy, whilst those leading from a large bed-sore on the opposite buttock were found by Mr. Beck to be plugged, the thrombosis extending into the internal iliac vein on that side whence the emboli were projecting into the common iliac vein, and thence had been washed into the circulation.

It is a remarkable fact that acute pyæmia never appears to occur in single isolated cases, but invariably in groups of two or three; not necessarily absolutely contemporaneous, but only separated by short intervals of time.

Is pyæmia contagious? The French surgeons generally believe that it is highly so. I have never seen an unequivocal instance of its spread in this way. This, however, may be accounted for by the fact that, whenever a case of pyæmia has declared itself in hospital, immediate steps have been taken to guard against contagion by isolating the patient and disinfecting the ward. We act, in fact, as if its contagion were proved, although we may not in reality be in possession of this proof. There certainly appears, however, to be an epidemic influence that favours its development. It is often coincident, though not invariably so, with outbreaks of erysipelas, both in the hospital and out of doors.

I have already said that it is necessary for the surgeon to consider the nature, rather than the number, of the cases contained in a ward. This is well exemplified by an outbreak of pyæmia that took place in Ward 1, in January 1871. In the early part of the month this ward contained a rather large number of severe wounds, including a sloughing cancer in the groin, an amputation of the leg, a compound fracture of the fibula with extensive laceration of the soft parts, a compound fracture of the femur, and one of the tibia.

Pyæmia occurred in the case of amputation of the leg on January 19; in the compound fracture of the fibula on February 3; in the compound fracture of the femur on the 13th: they all died. At no time, however, was the ward overcrowded as to number of patients, two out of the fourteen beds being empty the whole time; so that each patient had a cubic space of no less than 1,800 feet of air; but it contained for two months an average of seven or eight open wounds, of which five were always severe, such as amputations or compound fractures. During this period there was only one case of erysipelas, and the hospital generally was free from that disease. Another instance of a similar kind occurred in December 1872. At this period Ward 1 became again crowded with severe wounds, there being an average of nine open wounds, seven or eight of which were severe, including three amputations and an extensive necrosis of the tibia. One of the amputation cases had pyæmia, and recovered; another had pyæmia, and died. On this occasion erysipelas broke out in three cases. In both these outbreaks the weather was cold.

In erysipelas we find various influences tending to develop the disease—contagion, overcrowding, and epidemic influence all produce their effect. About the contagion of erysipelas there can be no question. I could adduce many instances of it, but one especially occurs to me. Though it happened many years ago, it has made a deep and lasting impression upon my mind. On January 17, 1851, a case of phlegmonous erysipelas of the leg was accidentally brought into No. 1 Ward. As soon as the nature of the disease was discovered the patient was removed to the erysipelas-ward, at the top of the building, having only remained in No. 1 for about two hours. At this time No. 1 Ward was perfectly healthy; but a few days afterwards a patient lying in the next bed to that into which the erysipelatous patient had been taken, and who had been operated upon for necrosis of the ilium, was seized with erysipelas. On the 22nd I performed five operations on patients who were in this ward. Of these cases three were attacked by erysipelas on the 24th, namely, a case of necrosis of the tibia, of partial amputation of the foot, and of encysted tumour of the scalp. All these patients died. On the 24th a patient was operated on for strangulated femoral hernia. He was seized with symptoms of low peritonitis, doubtless of erysipelatous character, on the 31st, and also died. No case of pyæmia showed itself, but all the patients in the ward who were not attacked by erysipelas had diarrhœa and severe gastro-intestinal irritation.

Here was a series of most lamentable catastrophes, doubtless directly induced by the accidental intrusion of an erysipelatous patient into an operation-ward, and then propagated by contagion.

That erysipelas may be developed in other ways there can be little question. I believe that by no means an uncommon cause is infection from dissecting students. I have several times seen cases, both in hospital and private practice, distinctly referable to contamination from this cause; and I think that no student, during the time that he is engaged in dissection, should be allowed to serve as a dresser in an hospital.

Another frequent cause of the infection of wounds in hospitals is the practice of allowing the house-surgeons and dressers, whose business it is to attend to the living, also to make the inspection of the bodies of the dead patients. In the exercise of this double duty they may go direct from the dead-house, where possibly has been conducted the examination of the body of a patient who has but a few hours previously died of the most infectious form of hospital disease, of pyæmia, erysipelas, diffuse peritonitis or septicæmia, into a ward full of patients suffering from wounds the result of injury or of operation. Who can be surprised at the atmosphere of the ward becoming contaminated and the wounds becoming infected by the most noxious of animal poisons, when they are dressed by the very hands which have recently conducted a post-mortem, or recently been immersed in the fluids of a body dead of infectious disease? The practice of allowing the same officer to discharge at the same time duties that are so antagonistic in a sanitary point of view as those in the ward, the operating-theatre, and the dead-house, is most reprehensible. It is undoubtedly a fertile and often an unsuspected cause of the propagation of disease in hospitals. But it is not to hospitals alone that this pernicious practice is confined. I have more than once seen erysipelas develop itself in private patients from this cause. Surgeons have much to learn from obstetricians in the prevention of infection. No accoucheur who has the slightest regard for the safety of his patient will attend a woman in her confinement if he has recently been exposed to the infection of puerperal peritonitis, or even of some of the ordinary forms of zymotic disease; and traumatic erysipelas certainly stands in much the same relation to surgical wounds that infectious peritonitis does to parturition; and yet how rarely is this obvious truth acted upon!

There is a remarkable resemblance, if not an actual similarity, between puerperal peritonitis and the erysipelatous inflammation of the peritoneum following operations on the organs contained in that cavity. The influence of contamination in the way that I mention is well illustrated by what occurred some years ago at the Vienna Hospital.

In the year 1839 the Maternity of Vienna was divided into two clinics—one for midwives, the other for students. In the students' clinic the mortality increased to a formidable extent, and up to June 1847 it was the seat of a murderous endemic (*endemie meurtrière*). During this period of eight years and a half the mortality amounted to 10·4 per cent., and this high rate was probably below the mark, for, according to Spath, those patients who were transferred to the general hospital and died there were not included in it.

In the midwives' clinic the total mortality during the whole of this period only amounted to 3·8 per cent., showing a mean difference of more than 6 per cent. between the results of the two clinics. This difference could not fail to attract the serious attention of the physicians, and Semelweiss, who had charge of the students' clinic, attributed the mortality there to the propagation of puerperal fever by septic matters.

In May 1847 he took every possible precaution against this source of contagion. The mortality diminished in a notable manner: it fell below the mean of the midwives' clinic, and since that time the destructive epidemics that used formerly to occur have almost entirely disappeared. In the two last years the results have been very remarkable, the mean mortality barely reaching 2 in 100. Amongst the means that he adopted were especially these: that no pupil attending women in their delivery was allowed to assist at post-mortem examinations of puerperal cases, and they were all compelled to wash their hands frequently in chlorinated solutions after making vaginal and other examinations. The healthy condition of a maternity charity, just, indeed, as that of a surgical hospital, depends in a great measure upon the precautions which are taken by all the persons attached to it in matters of personal cleanliness as well as in the arrangement of the wards and distribution of the patients.[1]

That erysipelas is often of epidemic origin there can be no question; but the influence of any epidemic is immensely increased by an unhealthy condition of a ward from overcrowding. Of this I will give one striking instance.

In the middle of November 1872, Ward 1 was filled by a very severe set of cases. There was one patient who had both legs amputated below the knees; another whose leg had been amputated through the knee; another who had a compound fracture into the ankle-joint; one with extensive necrosis of the tibia; and another with extensive laceration of the fore-arm. And this condition of severe cases was kept up by two other primary amputations—one of the foot, the other of the thigh—these cases being admitted in December, but whilst the preceding ones were under treatment. There was no erysipelas in the ward until November 15.

[1] Lefort: 'Des Maternités.' Paris, 1866, Pp. 151, 152.

On that day the case of compound fracture into the ankle was attacked. He recovered, and the erysipelas had left him by the 21st. On this day the patient with necrosis of the tibia had the sequestrum removed. He was a feeble old man. He was attacked by erysipelas on December 1, was removed to the erysipelas ward, and died there on the 7th. On December 4 the patient with scalp-wound was attacked, and was at once removed to the erysipelas ward. During this period, viz. on December 7 and on the 22nd, two of the other patients—the one with amputation through the knee-joint, and another—were seized with pyæmia. The patient who had been amputated through the knee, was the only one out of the twenty-three patients with acute traumatic pyæmia who recovered. During the same period two of my patients in Ward 5, both suffering from chronic wounds—one a fistula *in ano*, the other a sinus in the thigh—were attacked by erysipelas. To them it is possible that the infection was conveyed by the house-surgeon or dressers; but, as other cases of erysipelas occurred in the hospital at the same time, and one was admitted from out of doors, it is also possible that an epidemic influence existed which led to these attacks.

LECTURE IV.

ON THE PREVENTION OF HOSPITALISM.

In the last Lecture I pointed out to you that, and in what way, overcrowding gave rise to septic disease; and I particularly endeavoured to impress upon you the very important fact, which I cannot too strongly urge upon your attention, that it is the nature, and not the number, of the cases in a ward that occasions the pestilential state of its atmosphere which develops pyæmia and other septic diseases. A ward may be 'overcrowded' with those surgical injuries and diseases in which there is no breach of surface—such as simple fractures, chronically inflamed joints, strictures, &c.— without any more serious detriment to the health of its inmates than would result from inhabiting a room that is habitually too full of people; but the case is widely different when the inmates are the subjects of injuries or operations that occasion suppurating wounds. Then, and under such circumstances, even if the ward be not full, septic disease may be developed; and this will certainly happen, so far as our experience here will lead us to determine any sanitary point with certainty, if the number of suppurating wounds, the majority being severe ones, exceeds one-half of the number of patients the ward is constructed to hold; and further, when once septic disease is so generated, it will spread by contact or infection of air.

Let me, then, advise you not to be sceptics about the influence of septics. How this 'overcrowding' generates septic disease in the first instance—by local infection or by blood-poisoning, or by both—are points on which men of science are divided. The discussion of this question would be altogether beyond the object of these Lectures, and it need, consequently, not detain us. Its solution one way or the other in no way influences: the fact with regard to the evil influence of overcrowding. We may, then, take it as proved, by the experience derived from this hospital, that under circumstances of injury or disease, and of a building such as we have to deal with here, an invasion of septic surgical disease may be foretold.

We are able to determine with accuracy when it is imminent, and with certainty that it will occur according to the number and nature of the open wounds in a state of suppuration that are in a ward at a given time. To be forewarned in such cases is truly to be forearmed. As septic disease, and pyæmia more particularly, is the main cause of the present high rate of mortality after operations, more especially amputations; as the conditions under which it is developed can be determined with accuracy, we have only to guard against the development of these conditions in order to prevent the evolution of the septic cause of mortality, and thus, in the same ratio, diminish that mortality.

There are probably three essential conditions that primarily and chiefly influence the rate of mortality in any given hospital. These are—

1. Its size—measured by number of beds.
2. The amount of work done.
3. The mode of construction.

1. As to size of hospital, as counted by number of beds. Dr. Farr, the able and learned Superintendent of Vital Statistics, at the Registrar-General's Office, constructed the following Table from official documents. It certainly shows clearly the influence of mere size, irrespective of all other circumstances, on general hospital mortality:—

Principal General Hospitals in England and Wales, 1861.[1]

(Special Hospitals are excluded from this Table.)

	Number of Hospitals	Inmates	Average Number of Inmates in each Hospital	Deaths	Mortality per cent.
Total Hospitals .	80	8535	107	6220	72·88
Hospitals containing 300 Inmates and upwards .	5	2090	418	2101	100·53
200 and under 300	4	913	239	838	91·78
100 and under 200	22	2898	132	2041	70·43
Under 100 . .	49	2634	54	1240	47·08

Dr. Farr writes as follows: 'In the meantime it is evident that the mortality of the sick who are treated in the large general hospitals of large towns is twice as great as the mortality of the sick who are treated in small hospitals in small towns. It remains to be seen

[1] 'Twenty-fourth Annual Report.' 1863. P. 230.

whether the mortality in small hospitals is not twice as great as the mortality of the same diseases in patients who are treated in clean cottages. Should this turn out to be the case, the means of realising the advantages of the *hospital system*, without its disadvantages, will then be sought, and probably found, as the problem is not insoluble.'[1] Simpson endeavoured to prove that the mortality from septic disease in a hospital increases in the direct proportion of the size of the building. In this he was in error. It is quite possible that, if the conditions were equal in all hospitals, size might determine result; but the conditions are so unequal that no legitimate deduction as to mortality can be drawn merely from the consideration of the size of a hospital. If all hospitals were constructed in exactly the same way, if the number of severe injuries and operation-cases they contained was in exact proportion to the general number of their inmates, mere size of building, or, in other words, accumulation of numbers under one roof, would, in all probability exercise a determinable influence on the results. But, in point of fact, hospitals differ immensely in these respects, and the largest hospital in London, St. Bartholomew's, is at present, surgically, the most healthy. Hence, mere size cannot be considered as the sole cause in looking to differences of mortality.

2. The next condition that may influence result is the amount of work done in different hospitals. There can be little doubt that the work that is done in most hospitals at the present day is very far in excess of that for which they were originally intended by the con-

[1] 'Twenty-fourth Annual Report of the Registrar-General,' p. 231.

structors. Town populations, especially in the artisan and labouring classes, have immensely increased during the last half-century. The introduction of machinery, the crowded traffic of all great thoroughfares, and the vast amount and dangerous character of the work done at railway stations, docks, and other similar establishments, have greatly increased the number of serious accidents amongst the labouring class. Added to this the great spread of disease by hereditary transmission, increasing as it does in an ever-widening circle, affords an easy explanation of the great increase in surgical cases of a heavy character in all our hospitals.

But although the amount of surgical work done has enormously increased in most of the metropolitan and provincial hospitals, it varies to a far greater extent than is, I believe, generally known in different institutions, not only absolutely but relatively to the number of beds in different hospitals; being at least twice as great in proportion to the number of patients in some hospitals in London as it is in others; so that no safe deduction can be drawn from mere size of building or number of inmates. Thus, if we take the four hospitals in Table B, and add University College Hospital, I find that the number of cases of amputation in these hospitals in proportion to the number of beds they contain is as nearly as possible as follows, viz. :—

Hospital A, there was 1 amputation to every . . 13 beds.
Hospital B, ,, 1 ,, . . 17 ,,
Hospital C, ,, 1 ,, . . 10 ,,
Hospital D, ,, 1 ,, . . 35 ,,
University Coll. Hos., 1 ,, . . $6\frac{1}{2}$,,

At University College Hospital the number of operation cases has always been very large in proportion to the size of the institution. We have 150 beds, of which about 75 are surgical. Now, I find by the Registrar's returns that we have a yearly average of capital operations—*i.e.* of operations by which life is directly imperilled, as the major amputations, resections, operations for stone, hernia, the removal of large tumours, of organs, such as breast, testes, penis, and tongue— amounting to about 84, or more than one for each surgical bed; and this exclusive of all injuries of the head, chest and abdomen, and of all fractures that do not require amputation (*Vide* Table C). This is more than double of what occurs in some other hospitals in London of the same size, and fully as many as occur in some that contain more than double the number of beds. Hence in University College Hospital the proportionate number of severe cases in regard to the actual number of beds is very large, and yet the mortality from septic disease is, upon the whole, small.

In speaking so much of University College Hospital, I may possibly have laid myself open to something like a charge of egotism, for we naturally connect the hospital with the surgeons attached to it; but I must either refer to the results of the practice of that hospital, to which I have for nearly a quarter of a century been one of the surgeons, and from which I necessarily derive all my experience of public surgical practice, or I must be silent.

TABLE C.—*Showing number and nature of all Major Operations performed at University College Hospital in the Three Years, 1871-72 and 1873.*

Nature of Operation	Number of Cases
Major amputations of limbs	78
Resections of six larger joints	19
Lithotomy and lithotrity	37
Strangulated hernia	23
Cancer of breast	24
Other tumours of breast	6
Large tumours from various parts	27
Necrosis of jaws and excision of shafts of long bones	15
Removal of jaw	3
Do. tongue	6
Do. penis	3
Do. testis	4
Colotomy	5
Ununited fracture	2
Ovariotomy	1
Total	253

Being a yearly average of 84.

This table does not include any of the minor operations, which are very numerous—such as partial amputations of the hand and foot, however extensive, provided they were not through the wrist or ankle, thirty-six in number; tracheotom caries, operations for deformities, congenital or acquired; for stricture, perineal and other fistulæ, diseases of rectum, &c.; nor any operation occurring in the obstetric or ophthalmic departments.

The amount of active surgical operative work thus done in University College Hospital is very large in

FAULTY CONSTRUCTION OF HOSPITALS. 91

proportion to its size, *i.e.* the number of beds it contains; and yet the average mortality, as measured by the amputative death-rate, is small. Neither mere size of nor surgical activity in a hospital can thus be taken as the cause of the high rate of mortality that sometimes prevails. We must, therefore, consider the bearing upon this point of the last of the three conditions referred to, viz. hospital construction.

3. It is not my intention on the present occasion to enter upon the general question of the construction and hygiene of hospitals; but there are a few points to which, as they bear on our present subject, I may direct your attention; for, indeed, I cannot but come to the conclusion that in the solution of this question is involved much that relates to the development of septic disease, and that increases immensely the evil results which are consequent upon overcrowding, or the simple aggregation of patients under one roof, by rendering the air impure before it enters the wards, and keeping, in fact, the whole hospital atmosphere in a state of pollution.

I cannot but think that there is something radically wrong in the conventional method of constructing ordinary hospitals. I am not at present speaking of those great endowed hospitals founded by the munificence of the charitable and pious of a long past age, the revenues of which have augmented by the increased and ever-growing value of property; by the bequests and donations of successive generations of benefactors, until they have rivalled the incomes of dukedoms; but I solely speak of that large class of hospitals (metropolitan and county), varying in size from 100 to 300 or

400 beds, which have no such princely revenues to fall back upon, but are supported from year to year by the 'voluntary contributions' of the many benevolent, charitable, and often munificent, donors who live amongst us. These hospitals, having often but slender means at their disposal, or having been constructed with economy in outlay, have been built on a conventional plan, from which no architect has yet departed. They are simply big houses, with basements containing kitchens, sculleries, cellars, and the ordinary offices of a large establishment; with an operating theatre and dead-house more or less closely connected with the main building; with every floor filled with sick and injured people. On the ground-floor, accidents and operation-cases; on the first-floor, probably medical patients; above, chronic surgical cases—who can wonder at the development of pyæmia below and of erysipelas above? Who would live in an ordinary house thus filled? Who would expect to preserve his health if he ventured to inhabit it? How can the air be pure, and how can recovery be expected, or freedom from septic disease secured, under such circumstances? Civil hospitals, that were built during what may be termed the 'pre-sanitary age'—*i.e.* until about a third of a century ago, when 'drainage' and 'sewage' had not become topics of drawing-room conversation, when such a being as a professor of hygiene was yet undreamt of—were uniformly constructed on this 'big-house' plan, being three or four storeys high. The evil of this mode of construction is very great. It leads to the upper storey, which ought to be the healthiest, being usually the most infected with septic disease, and this notwith-

standing its being more easily ventilated than those lower down.

This has happened here repeatedly with regard to erysipelas. The upper floor in the main building of our hospital is occupied by the female surgical beds. The cases are much less severe than those in the male accident wards on the ground-floor; the wounds are slighter; operations less frequent, and far less severe. But yet erysipelas occurs to a greater extent amongst the patients on this floor than amongst those on the ground-floor. From July 1870 to July 1873 there were thirty cases of erysipelas in these wards, against twenty-four in the lower wards. But during the whole of this time I had not one single case of pyæmia amongst my female patients in the upper ward. The fact appears to be, that the air, as it ascends in the hospital, becomes impregnated more and more with septic matters; and that, erysipelas being infectious, the contagion of that disease is at once carried by the upward current of air to the higher storeys, however well-ventilated these wards may be; and no doubt the good ventilation in these higher wards tends to lessen the amount of septic disease which would otherwise be more frequent in them.

But not only is the general mode of construction of most hospitals, whether town or provincial, faulty, but the internal arrangements are usually such as not only tend to predispose to those septic diseases that are due directly to contamination of air beyond a certain point of impurity, but also to render it extremely difficult in all cases, and in some impossible, to expel or eradicate such disease when once it has got a foothold in the building. A hospital may thus become contaminated by septic disease—be pyæmia-stricken, in fact—beyond the possibility of purification. This is what has happened at the Lincoln County Hospital, which the governors have most nobly decided on demolishing and reconstructing at a vast cost. In the case of this hospital, to which I have no hesitation in alluding, as its condition has been made public in the local papers, in the medical journals, and, at the meeting at which its reconstruction was decided upon, my name was mentioned in reference to a modified scheme, there have been repeated outbreaks of pyæmia for many years past; and at last the surgeons had been compelled to desist from operating in it, and even to advise patients to be treated at their own homes, rather than encounter the perils of the hospital. Mr. Brook, one of the surgeons, said, at the meeting alluded to: 'Of late years the interior of the hospital has, at one time or another, been entirely renewed; but still the disease (pyæmia) kept breaking out; and it was the opinion of all great authorities that it lurked in the very fabric, and that nothing but demolition would remove it.' ('Lincoln Gazette,' January 10, 1874.) Mr. Lowe, another of the surgeons, spoke in the same strain, and has told me that, although the hospital had become thus pyæmia-stricken, he had never met with a case of that disease out of the hospital in the town itself. Mr. Sympson, another of the surgeons, gives equally strong expression of opinion as to the necessity for complete reconstruction, in an extremely interesting pamphlet on the subject.[1] He says: 'The effect of the shortcomings of the

---

[1] 'Letters concerning the Lincoln County Hospital,' by Thomas Sympson. Lincoln, 1873.

hospital is to occasion extreme risk to patients in the event of a successive occupancy of the beds by urgent and offensive cases.' (P. 11.)

The evidence of Mr. Cadge, the able and experienced surgeon of the Norwich and Norfolk Hospital, is equally conclusive on this point. In that hospital, which was the surgical home of Gooch and of Martineau, of Crosse and of the Dalrymples, pyæmia has become so rife as to be a most serious obstacle to successful operations, whilst out of the hospital, in the town and surrounding country district, it is unknown. Mr. Cadge says: 'It is now about twenty years since first I became attached to the Norfolk and Norwich Hospital. I am grieved to say that for a considerable part of that period the building has fallen upon evil times with regard to this question of pyæmia. We have had within the last six or eight years a decidedly increasing amount of pyæmia in hospital practice, and that notwithstanding that the hospital is fairly well-provided in regard to cubic space and beds, with a perfect system of drainage which has been established within the last few years, and with other precautions which have in times gone by seemed to secure us from these plagues of surgery. Still the fact remains, that during the last three years—I go no further than that—we have had twenty-one or twenty-two deaths from pyæmia in the hospital. Yet this is on the whole clearly and thoroughly to be accounted for. The way of accounting for it seems to me to touch the very question we have at issue between private and hospital practice; and it arises chiefly, in great part certainly, from overcrowding the wards of the hospital. During twenty, or twenty-five, or thirty years we have had no extension of the hospital. It accommodates about 140 patients, 70 or 75 on each side. We have had no extension, but during that period there has been a large increase of population, and there has been a still greater increase in the growth of mechanical contrivances and machinery, particularly of agricultural machinery, which has thrown into the hospital a very large increase of serious injuries, accidents, and wounds of all kinds. I have no doubt in my own mind that the increase and development of pyæmia in that institution is largely due to this increase of serious accidents and wounds, which have not been properly cared for by extension of hospital accommodation. Some proof of this is found, I think, in this fact. Of course by far the major part of these cases of injury and disease, and of all other surgical operations and diseases, occurs on the men's side of the hospital. Out of twenty-one cases of pyæmia during the last three years, only one has occurred on the women's side, and this, I think, must be a proof that we have an overcrowding, not in the sense of a deficient cubic space, but in the sense of numerous cases of a bad kind in that space which we could not well avoid. We have been obliged to take serious steps towards altering this condition of things, and those steps are now in progress. Still that state of things remains. We have shut up one ward after another, we have abolished the use of sponges, we have weeded out beds from each ward, we have looked after the ventilation in a very careful way, and the result of these experiments we have not yet had time to prove. So far as we can tell, I am afraid that they have wrought no benefit, and I should not wonder that the end is that we either close the hospital or make a large extension or a full reconstruction of it. Well, if that be the case in hospital practice, I am happy to turn to the absence of pyæmia so far as private practice is concerned. I may be permitted to say that during the last fifteen or twenty years a considerable share of both surgical operations and all kinds of surgical cases has fallen in my way in the city and in the country for twenty or thirty miles round, and I think I can say that during twenty years and more, with one exception, I have not come across a case of pyæmia; and I am not sure that that one exception can be regarded as a typical unequivocal case. Neither have I seen any case following from carbuncles or boils. In this way I have been led to think that one's private practice was perfectly secure and very much opposed to hospital practice. I have unwillingly and almost tremblingly proceeded to operate in the hospital, but I have had a happy confidence and a perfect assurance that in all private cases I should avoid any of these disastrous consequences.' And he goes on to make this pertinent remark: 'I come to the conclusion in my own mind that pyæmia, if it does not find its birth-place, does find its natural home and resting-place in hospitals; and although a hospital may not be the mother of pyæmia, it is its nurse.'[1]

Such evidence as this, coming from a provincial surgeon of the highest standing and largest experience, is peculiarly valuable, as showing that in provincial towns there may be the same difference, so far as liability to pyæmia is concerned, in the hospital and in

[1] 'Lancet,' vol. i., 1874. P. 338.

the private practice of the same surgeon, as is found by many to exist in London.

I am acquainted with other instances of hospital infection nearly, if not quite, as serious as this; but, as they have not yet been brought before the public, I refrain from mentioning them. But surely no stronger or more conclusive evidence is needed of the tenacious and ineradicable nature of this pyæmic infection, when once it has taken firm hold of a hospital, than that which is furnished by the examples just given. What name so appropriate as 'hospitalism' for a condition of things such as here described? The town free from infection; the hospital saturated by it, to such an extent as to induce its own surgeons to recommend their patients not to enter it, to compel them to refrain from operating, and, after every attempt that science and humanity could suggest—every hygienic means employed in vain in the fruitless attempt to eradicate the pestilence from 'the very fabric' itself—to cause the governors, as a last resource, to decide on the demolition of the building and its complete reconstruction, at a great expense, as the only remedy. The truth is that, when once a hospital has become incurably pyæmia-stricken, it is as impossible to disinfect it by any known hygienic means, as it would be to 'disinfect' a crumbling wall of the ants that have taken possession of it, or an old cheese of the maggots which have been generated in it. There is, in these extreme cases, only one remedy left—that remedy which the governors and staff of the Lincoln County Hospital have so generously, so disinterestedly, so nobly resolved on—viz. the demolition of the infected fabric, and we must add the destruction of its materials, for that last clause must not be forgotten. In fact, just as the cattle-plague has to be 'stamped out' by the pole-axe, so has the infection of a pyæmic hospital to be destroyed by the pick.

From the manner in which most hospitals are built —on the model, in fact, of a large dwelling-house—all the wards open upon corridors or lobbies, which communicate with the general staircases of the building, and, through the medium of these, with its basement-floor, with the entrance-hall, the out-patient department, and frequently by continuous passages with the dead-house. The only air-supply that the wards receive at night, or when the windows are shut, is, in the majority of cases, that which is thus drawn from these corridors, and which, from various causes, is already in a state of impurity, and will thus more readily become contaminated up to the septic point by overcrowding of the wards.

Now, let us consider briefly how these causes of pollution of the hospital air may be remedied.

The three main causes of the impurity of the air of hospital corridors and staircases appear to be: 1. Effluvia from the kitchens, cellars, washing-places, sculleries, and dust-bins, on the basement floor. Mr. C. de Morgan has related a very interesting and important fact in reference to this point. In the Middlesex Hospital the patients who occupied two beds, one on either side of a particular window, were peculiarly liable to erysipelas and pyæmia. It was found that there was a dust-bin in the area below, in the direct line of the window. This was cleared out and disused.

The erysipelas and pyæmia disappeared from the two beds. After two or three years the dust-bin was again used, and the septic diseases again made their appearance in the occupants of these two beds. The odours arising from the various sources of impurity just mentioned, the smell of cooking, of washing, &c., are often perceptible in a very marked degree in the lower floor of hospitals. This evil might easily be remedied by removing all these offices into an outbuilding, as is now generally done in well-constructed modern dwelling-houses. The basement-floor might then be converted to useful purposes for the inmates. Reading and recreation rooms might be established in it for the use of residents or patients.

2. The second great cause of impurity in the internal air of many hospitals arises from the out-patient department being under the general roof of the hospital. This I cannot but consider to be a great evil, and a most fertile source of disease amongst the inmates. The fact is, that there is too much done in hospitals at the present day. They have not been constructed originally for the amount of work that is thrown upon them. Thirty years ago, the out-patient department was insignificant in comparison to what it now is. Dispensaries then did the work that hospitals now undertake in this respect. Not only are there the usual general medical and surgical out-patients, but very commonly out-patient arrangements are made for those affected by a variety of special diseases. Not only has the augmentation in the number of out-patients become so great as to be a source of demoralisation to the public at large, of loss and injustice to the great

mass of medical practitioners, of wasteful expenditure and of serious embarrassment to hospitals, but the accumulation of these crowds of diseased, often of infectious, people in the entrance-halls and 'out-patient and casualty' rooms of hospitals, has become a source of serious unhealthiness to the inmates of these institutions. The out-patient room is commonly situated near one of the entrances into a hospital; and who that has gone into that apartment, when crowded with patients, many afflicted alike with dirt and disease, has not been conscious of a heavy and noisome odour tainting the air at its first entry into the hospital, and rendering it, at the very doors of the building, unfitted for contact with the wounded within its walls?

I give no opinion here on this question of indiscriminate out-patient relief. Whether an evil or a necessity, it exists. But it should not be suffered to exist under the same roof that covers those who have recently sustained serious injuries or been subjected to grave operations. For this, at all events, there can be no necessity; and we shall in vain attempt to purify the air of our hospitals, and to render it free from septic influences within, while we allow it to be polluted, if not infected, at its very entry into the building. Let the out-patient department be removed from the hospital to a detached and altogether separate building, and we shall at all events remove one source, and a great one, of atmospheric impurity.

3. A third grave cause of the impurity of the *general* as distinguished from the *ward* air of hospitals, is often the proximity of the dead-house and *post-mortem* room to the main building, with which it, in most instances,

is connected by means of a corridor. To this I need do little more than allude; for it must be obvious that it is hopeless to endeavour to prevent the continuance, the occasional severe outbreak, or the spread, of septic diseases in the wards, when the air that supplies them from within the building is from time to time contaminated by the diffusion through it of emanations arising from the *post-mortem* examination of those who have already perished from their influence. The dead-house, like the out-patient department, should be entirely separated from the hospital, and should have no covered communication with the main building. So also the porters whose duties are connected with it, and the officers whose business it is to make the necropsies, should on no account be permitted to handle the patients or to examine or dress their wounds.

There is one other source of infection, or rather of maintenance of infection, in a hospital that is closely allied in its influence to the dead-house; I mean the 'erysipelas ward.' That, also, should never be allowed within the walls and under the roof of the main building, but should be detached and non-communicating. The contagion of erysipelas is undoubted. I gave you, in the last Lecture, a striking illustration of it. What but contamination, then, can be expected from placing patients so diseased on the same floor, in the same stream of air-communication with others, rendered, by open wounds, peculiarly susceptible to its infection? If the 'erysipelas ward' were removed out of the hospital, and had no communication with the main building, its use would be greatly restricted; as now situated, it too often feeds itself.

INFLUENCE OF AGE OF HOSPITAL.

We know as yet but little that is positive as to the influence of the duration of use, or, in other words, the age, of a hospital in producing a tendency to the generation of septic diseases within it. It is a question extremely difficult to answer, how far the mere age of a building tends to its unhealthiness—whether that building be a modern hospital or a mediæval Italian palazzo. There are so many circumstances in every building, independently of its mere age, that must modify its sanitary condition—such as the material of which it is constructed—stone, brick, or modern lath-and-plaster—its site, and the attention that has been bestowed upon its drainage and other hygienic conditions—that we cannot, I think, go much further than this in the answer, viz. that age and neglect will certainly render any building unhealthy.

There is one remarkable circumstance connected with age of hospital, and it is this, that new buildings added on to old hospitals often become more unhealthy than the original building. This happened at one of the Paris hospitals (I believe it was the Neckar), and has been several times observed in this country. It is difficult to explain this fact—for a fact it is. Whether the new materials of a recently-built edifice are more absorbent than those of an older one, and thus easily become impregnated with the impurities of the polluted air that enters from the old contiguous building, or whether the glue and size used in the construction of the building, being animal matters often in a state of semi-decomposition when used, undergo a septic change, and thus become actively noxious, remains to be determined. But any way experience has shown it to be unwise in a sanitary point of view to add new hospital buildings on to old ones.

But to return again to that on which I base all my remarks in these Lectures, viz. the experience I have derived during the twenty-four years that I have been surgeon to University College Hospital, I can say this, that—notwithstanding the immensely improved hygienic arrangements that have, during that time, been introduced and adopted by the unceasing vigilance and devoted attention of the hospital committee—the sanitary state of the hospital has just been maintained at the same point that it was a quarter of a century ago. Our amputation-mortality is exactly the same now as it was then, viz. about 25 per cent. The effect of greatly improved hygienic arrangements has been to prevent the hospital from becoming more unhealthy; they have not succeeded in increasing its salubrity. They appear just to have been able to counteract the ill effects that would otherwise have arisen from the impregnation of the building with the accumulated septic emanations of a continuous influx of sick and wounded. With regard to other hospitals, the same remark probably holds good; for, as I showed in the first Lecture that their amputation-mortality has not decreased, it may be presumed that their sanitary condition has not improved during this time.

There are various other points' in connection with hospital hygiene to which I need do little more than direct your attention, as they are matters of general rather than of special sanitary arrangements; they are matters, in fact, of common sense and of common experience. Though generally known, they are often neglected, but attention to them will undoubtedly materially lessen the development, the prevalence, and the persistence of septic disease in a hospital. They consist mainly in the following points:—1. The rendering of all surfaces as little as possible absorbent, and retentive of septic emanations, by parianising the walls and painting the floors; by these means also they become smooth and less likely to harbour organic dust. 2. The employment of dry rubbing of, instead of washing, the floors. 3. The frequent purification of the bedding—the blankets more especially, which are often very imperfectly cleaned and purified, and, like all woollen fabrics, harbour infection long and tenaciously. 4. The removal of all unnecessary furniture, such as bed-curtains, carpets, &c., that can impede ventilation or harbour infection. 5. The removal of the patients' clothes, especially those of cloth and woollen materials, from the wards, and storing them away in an out-building. 6. To compel the nurses to wear dresses that can easily be, and that frequently are, washed. 7. It would doubtless be advantageous to furnish the patients with hospital suits of clothes that admit of being cleaned and washed—flannel in winter, cotton in summer. 8. There should always be an abundant supply of carbolised water in the wards for washing purposes. 9. No sponges should ever under any pretence be allowed in the wards; those employed in the operating theatre for recent wounds should be very frequently renewed, and after use soaked in strong carbolic acid solution. 10. No personal communication through porters, dressers, or house-surgeons should be permitted between the dissecting-room, or dead-house, and the wards. 11. The isolation of patients suffering from old or fœtid ulcerations, more especially those of a cancerous character. 12. The separation of patients with suppurating wounds from one another by the interposition of unwounded patients. 13. Care not to allow more than one-half of the patients in any given ward to have suppurating wounds, even if these wounds be trivial; nor more, if possible, than one-third if severe. 14. Instantly to isolate all cases of septic disease. 15. To see that there is a current of air admitted into and passing through the ward from the outside of the building, by night as well as by day. 16. To close every surgical ward once a year for a month, during which time it should be disinfected and whitewashed. 17. And above all, and under all circumstances, to avoid overcrowding, for however short a time.

By attention to simple hygienic rules such as these much may be done to lessen the mortality from septic surgical diseases in hospitals; to mitigate the violence of their outbreaks; to limit the range of their destructiveness when they do occur; and, above all, to prevent the risk of ineradicable impregnation with and persistent contamination of the building by septic poison, a condition which experience has shown to be incurable, and only to be met by the destruction of the infected building. In fact, we may lessen materially, if not entirely remove, those external conditions which, independently of any predisposing constitutional state or purely epidemic influence, give rise to septic diseases which are truly the plagues of hospitals.

And now, Gentlemen, I have done for the present with this subject, the most important, probably, that is at this time before the profession. It is alike important to the public and to the surgeon. To the public, for whose use the hospitals have been constructed and are maintained, there can be no subject of deeper or more vital moment than that these hospitals should be kept in the highest state of sanitary efficiency. To the surgeon the importance of the prevention of hospital diseases cannot be over-estimated. As matters now stand, the most consummate skill and the most devoted attention are often alike rendered unavailing by the influence of septic diseases generated within the hospitals themselves by the operation of conditions entirely beyond the surgeon's control, yet admitting of prevention or removal—diseases which neutralise the best-directed efforts of his art, and which in no small degree increase the pressure of that heavy responsibility and deep anxiety which, even under the most favourable circumstances, attend its exercise.

www.ingramcontent.com/pod-product-compliance
Lightning Source LLC
Chambersburg PA
CBHW021943160426
43195CB00011B/1207